Allied Health Series

Cytology

Introduction to Cellular Structure and Function

Edmund J. Messina, Jr.

Consultant/Reviewer
Kola Kennedy, C.T. (ASCP)

The Bobbs-Merrill Company, Inc.
Indianapolis

FIRST EDITION

FIRST PRINTING—1975

The Bobbs-Merrill Company, Inc.
4300 West 62nd Street
Indianapolis, Indiana 46268

Library of Congress Catalog Card Number: 74-79836
ISBN: 0-672-61382-4

Preface

One of the most important developments in the health field in recent years has been the rise of allied health personnel. These persons provide essential services and support for physicians and dentists (including those engaged in specialized practices in these professions) as well as for others engaged in health-oriented fields, such as dental assistants, inhalation therapists, physical therapists, and medical records technicians.

Professional Needs

The day when the physician, in or out of the hospital environment, was able to carry the whole burden of patient care with no help other than that of a nurse is over. Today, the professional nurse herself needs well-trained assistants, and it is generally accepted that health care is a team operation, with the physician as the head of the team. Furthermore, each member of the team must be a skilled technician—able to carry out the increasingly complex techniques required by modern medical care.

Educational Needs

The pressing need for well-trained personnel prepared to serve in the various allied health occupations has opened a whole new field of education. It is no longer possible for a person interested in working as a technician in a medical, dental, or x-ray laboratory to acquire mastery of his chosen field by learning on the job. Although on-the-job training remains important, basic training in the classroom must precede it. This fact has become so evident that schools, laboratories, and hospitals and allied institutions have begun to step up their educational programs to meet the growing demand for trained personnel. Another factor that emphasizes the urgent need for academic training in the allied health fields is the demand by federal, state, and municipal regulatory agencies for the certification and licensing of personnel in an increasing number of these occupational fields.

Educators attempting to respond to the challenge of providing sound educational systems for developing personnel for these jobs found themselves handicapped by the lack of authoritative textbooks that met the special requirements of their students. Students interested in preparing themselves to fill these jobs found it difficult to obtain clear, concise information that was relevant. Persons already employed in allied health fields found it difficult to obtain the additional information they needed to augment and supplement their knowledge and experience.

In response to all of these needs, the Allied Health Occupations Series was conceived. The Series consists of textbooks, workbooks, audio tapes, laboratory manuals, and teacher's guides to provide a wide range of educational materials tailored to meet a variety of requirements and to serve students with various abilities.

About the Allied Health Occupations Series

The Allied Health Occupations Series is especially structured to offer a wide variety of basic tools for creative teaching and learning. Each component of the Series presents detailed, comprehensive information that is easy to read, easy to use, and up to date. All are specifically designed for maximum use by educators, students, and personnel in allied health occupations.

The modularized approach of the Series is unique. It allows unlimited flexibility for the teacher and for the student. The illustration on this page gives a few examples of the multiple use that can be made of materials because of the variance and overlap of informational needs in various allied health fields. Core materials (e.g., *Basic Chemistry* and *Basic Microbiology*) provide basic information; supplementary materials (e.g., *Organic Chemistry* and *Introduction to Diagnostic Microbiology*) provide specific and detailed information. Teachers and students can adapt the materials to meet their special needs.

The books have been carefully designed to invite learning and to aid teaching. Teachers will immediately notice the blocks of text material uninterrupted by hard-to-read italics or boldface type so popular in older textbooks where it was believed necessary to indicate "important" words or phrases. Such artificial techniques are unnecessary in these texts because each has been written to include only that information essential to a

MODULARIZED APPROACH

thorough understanding of the subject. Extraneous materials have been removed to provide the teacher and the student with basic concepts. In other words, all of the words and phrases are considered important for complete understanding of the subject.

However, it is easy to find important words and phrases or any kind of information in the books.

The Allied Health Occupations Series

Title	Format	Title	Format
Human Anatomy	Text	Orthopaedic Physician's	
Basic Chemistry	Text/Workbook	Assistant Techniques	Text
	Lab Manual	Respiratory Therapist	
Organic Chemistry	Text/Workbook	Manual	Worktext
	Lab Manual	Basic Microbiology	Text
Clinical Chemistry	Labtext	Introduction to Diagnostic	
Medical Office Practice	Worktext	Microbiology	Text
Cytology	Worktext	Medical Radiographic	
Basic Medical Laboratory		Technology	Worktext
Subjects	Worktext	Medical Mathematics	Worktext
Health Careers & Medical		Patient Care Techniques	Text
Sciences	Text	Dental Assistant Techniques	Worktext
Basic Medical Terminology	Workbook, Audio	Medical Records Technology	Worktext
	Teach Manual	Management Procedures	Worktext

In the front of each book a Table of Contents gives a general overview of each chapter, section-by-section. In the back of each book is a comprehensive Index that details information by word, by phrase, or by concept. Other built-in informational and instructional aids include charts, tables, drawings, and photographs that supplement and augment the text. Review questions help students as well as teachers judge the level of learning each step of the way, and suggested projects encourage students to practice and perfect professional skills. Specialized glossaries have been included where these have been deemed helpful for students.

All of the resources in the Series are generally interrelated to provide ease of use. Students and teachers need not readjust to completely different formats as they move from one subject to another. However, information is presented in a variety of styles to provide a change of pace that helps maintain interest.

About This Book

The book *Introduction to Cytology* is a combined textbook and workbook. It will serve as a valuable tool to supply various kinds of information about the structure and function of cells for persons in the Allied Health Occupations who wish to serve in the field of cytotechnology. The practicing cytotechnologist is primarily concerned with the effects of disease processes on individual cells. In order to recognize and to understand such disease processes, however, the cytotechnologist must first know the structure and function of the normal cell. This worktext, therefore, does not deal primarily with cellular pathology. Rather, it places the individual cells within the general context of the human body. It shows how the cells relate to the normal functions of the various organ systems of the human body, and how these systems and the cells that comprise them may be influenced by other systems, such as the endocrine system. The book is designed for the student who has a good background in biology.

Introduction to Cytology is profusely illustrated. The author has reduced the amount of text and provided illustrations, some of them quite simple, to carry the information. The illustrations are often presented in serial progression to highlight the points being discussed.

This worktext is intended to be used as a working tool for both the student and the instructor.

An introductory section that covers the fundamentals of cytology precedes the chapters on the individual systems. It provides the information that the student needs to understand the actual pathological processes of the body.

The author of this book is Edmund J. Messina, Jr., M.D./Ph.D. program, Claude-Bernard University, Lyons, France. Mr. Messina is a medical writer and translator and held the position of instructor of medical English at ADIT Institute in Lyons. He is enrolled at the University of Illinois Medical Center in a program to complete his degree requirements.

Kola Kennedy, who served as reviewer for this book, is a Certified Technologist (ASCP) and is Educational Coordinator, School of Cytotechnology. Michael Reese Hospital, Chicago, Illinois. She also serves as Chief Cytotechnologist in the Cytology Laboratory, Department of Pathology, Michael Reese Hospital, and has had several papers on cytology published in medical journals. In her review of this book, Ms. Kennedy states: "This text offers an introduction to human anatomy, physiology, and histology to the student who has some background in biology. The material is presented in a well-organized manner according to body systems, and the illustrations are simple and easily understood."

Instructional | Dynamics | Incorporated
Chicago, Illinois 60611

Philip Lewis, Ed.D.
President IDI

Linda J. Thomas
Editorial Director

Eleanor L. Bartha, Editor

Jane P. Barton, Associate Editor

Sophia A. Kaspar, Indexer

National Advisory Review Committee
Allied Health Occupations Series

Elijah Adams, M.D.
  Professor and Chairman
  Department of Biological Chemistry
  University of Maryland
  Lombard and Greene Streets
  Baltimore, Maryland

Robert O. Coddington, M.D.
  Director
  Orthopaedic Surgery Training Program
  Baroness Erlanger Hospital and
  T.C. Thompson Children's Hospital
  Chattanooga, Tennessee

Helen W. Dunn, R.N.
  Administrative Secretary
  State of Michigan
  Department of Licensing and Regulation
  Board of Nursing
  Lansing, Michigan

Webster S.S. Jee, Ph.D.
  Acting Chairman and Director
  Department of Anatomy
            and
Edward I. Hashimoto, M.D.
  Professor of Anatomy
  The University of Utah
  Salt Lake City, Utah

James A. Jackson, M.T. (ASCP), Ph.D.
  Technical Training Manager
  Technical Service Laboratory
  Ames Co., Division of Miles Laboratories, Inc.
  Elkhart, Indiana

Kola Kennedy, CT (ASCP)
  Educational Coordinator
  School of Cytotechnology
  Department of Pathology
  Michael Reese Hospital
  Chicago, Illinois

Leon J. LeBeau, Ph.D.
  Professor of Microbiology and Pathology
  University of Illinois at the Medical Center
  Department of Microbiology
  Chicago, Illinois

Robert G. Martinek, Ph.D.
  Chief
  Laboratory Improvement Section
  State of Illinois

Department of Public Health
Division of Laboratories
Chicago, Illinois

Edward Michaelson, M.D.
  Associate, Division of Pulmonary Diseases
  Mount Sinai Medical Center
  Miami Beach, Florida

Shirley Muehlenthaler, AAMA
  Program Chairman
  Medical Assistant Education
  Des Moines Area Community College
  Ankeny, Iowa

Eugene W. Rice, Ph.D.
  Clinical Biochemist
  The Allentown Hospital Association
  Department of Pathology
  Allentown, Pennsylvania

Joyce Sigmon, C.D.A.
  Director, Dental Assisting Education
  Council on Dental Education
  American Dental Association
  Chicago, Illinois

Carroll L. Spurling, M.D.
  Director, Regional Blood Program
  The American National Red Cross
  Los Angeles-Orange Counties
  Red Cross Blood Program
  Los Angeles, California

Murray Watnick, M.D.
  Radiologist
  Noble Hospital
  Westfield, Massachusetts

Ruth Williams, M. T. (ASCP)
  Professor and Chairman
  University of Florida
  Department of Medical Technology
  J. Hillis Miller Health Center
  Gainesville, Florida

Robert R. Zappacosta, RRA
  Director
  Medical Record Administrative Program
  Institute of Health Sciences
  Hunter College
  New York, New York

Contents

CHAPTER 1.   **The Cell: Structure**   9

Cell Membrane, 9; Cytoplasm, 10; Endoplasmic Reticulum, 10; Mitochondria, 11; Ribosomes, 12; Lysosomes, 12; Golgi Apparatus, 14; Cytoplasmic Inclusions, 14; The Centrioles, 15; The Nucleus, 15

CHAPTER 2.   **The Cell: Physiology**   18

Membrane Transport, 18; Active Transport, 19; Vacuole Transport, 20

CHAPTER 3.   **The Cell: Biochemistry**   23

Enzymes, 23; Energy, 23

CHAPTER 4.   **The Cell: Genetics**   26

Nucleic Acids, 26; DNA, 26; The Genetic Code, 28

CHAPTER 5.   **The Cell: Reproduction and Tissue Formation**   32

Chromosome Number, 32; Centrioles, 33; Mitosis, 33; Sex Chromatin, 35; The Cell Cycle, 35; Organization into Tissues, 35; Epithelium, 36

CHAPTER 6.   **The Female Genital Tract: The Ovaries**   40

The Ovary, 40; Follicular Development, 42; Endocrine Control of Ovulation, 45; Drug Controlled Ovulation, 45; Estrogens, 45; Progesterone, 45; Menarche, 49; Menopause, 50

CHAPTER 7.   **The Female Genital Tract: Fallopian Tubes and the Uterus**   59

The Fallopian Tube During the Endocrine Cycle, 60; The Uterus, 60; The Myometrium, 61; Physiology of the Uterus, 62; Uterine Contractions, 63; Menstruation, 63

CHAPTER 8.   **The Female Genital Tract: The Vagina**   71

Vaginal Flora, 72; Lactic Acid and Vaginal pH, 72; The Vagina and Sexual Activity, 72; The Vaginal Mucosa, 72; Section of the Vaginal Epithelium, 72; The Vaginal Epithelium Cycle, 75; The Vaginal Smear, 75; Cyclic Variations in the Smear, 79; The Endocervix, 80

CHAPTER 9.   **The Urinary Tract**   83

The Kidneys, 83; The Excretory Passages, 85; The Renal Pelvis and Ureters, 86; The Bladder, 87; The Urethra, 87

CHAPTER 10.   **The Male Genital Tract**   91

The Prostate, 94; Ejaculation, 96

CHAPTER 11.   **The Respiratory Tract**   100

Structure, 100; Histology, 101; Structure of Respiratory Tract Segments, 102

CHAPTER 12.   **The Digestive Tract**   107

Structure, 107; The Esophagus, 108; The Stomach, 108; The Small Intestine, 111; The Duodenum, 111; The Large Intestine, 113

RECOMMENDED READING   119

INDEX   120

chapter 1

The Cell: Structure

In order to understand the pathological changes in cells and how to detect them, it is essential that the cytologist have a thorough knowledge of cellular structure and function. This chapter will discuss the fine structure or ultrastructure of the cell and the specific biological function of each of its parts.

Cell Membrane

Fig. 1-1 is a schematic representation of an animal cell. The cell is bounded by the cell membrane (sometimes called the plasma membrane), which retains the contents of the cell. The cell membrane is believed to consist of a fatty or lipid layer sand-

wiched between two protein layers (see Fig. 1-1a). This membrane is an active, living structure—not just an inert container. The membrane appears to be continuous with other structures in the cell and contains a series of tiny pores on its surface. It performs many functions which are essential to survival.

A major function of the cell membrane is to keep certain substances out of the cell but to allow others to enter freely. This property is called selective permeability. The principle involved is much more complex than that of a simple gravel sieve because certain large molecules are allowed to pass freely but smaller molecules are prevented from entering. The mechanism is not fully under-

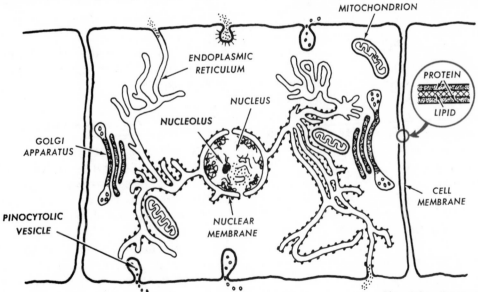

Fig. 1-1. Schematic representation of an animal cell.

Fig. 1-1a. Cross section of a cell membrane.

stood, but it does involve factors such as the electrical charge of the molecule and the molecule's ability to dissolve in lipids and then emerge on the other side of the membrane.

The cell membrane is known to vary in certain types of cells, such as the absorptive cells of the intestine. These cells need a large surface area to be efficient so the membrane in these cells is convoluted into small fingerlike projections called microvilli (Fig. 1-2). Another variation of the membrane is known as a desmosome, derived from the Greek word *desmos*, which means a bond. Desmosomes connect cells, such as epithelial cells, and create a strong cellular sheet.

protein fibers that link together to give gelatin its characteristic consistency. This is known as the gel state of a liquid and can be readily dissolved by heat or changes in pH. The dissolved gel then becomes a watery liquid known as the sol state. Thus, in a cell, there is a continual transformation from sol to gel and vice versa due to local changes as, for example, the acid substances produced by metabolism. In general, the peripheral hyaloplasm, or ectoplasm, is in the gel state and acts as a support for the cell membrane, while the endoplasm is more apt to be in the sol state.

The cell is an active place where substances constantly are being brought in, taken out,

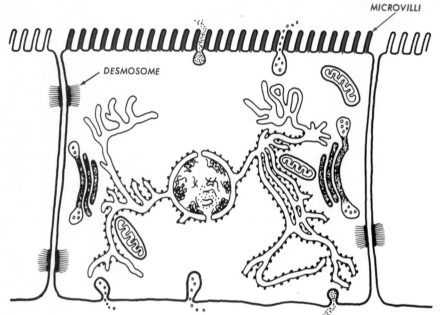

Fig. 1-2. Cell with microvilli.

Cytoplasm

The cytoplasm is the substance surrounded by the cell membrane. In it floats the nucleus and the smaller parts of the cell, known as the organelles, each of which has a specific function. The cytoplasm, which includes all the cellular contents except the nucleus, is sometimes divided into the peripheral cytoplasm, or ectoplasm, and the more central cytoplasm, or endoplasm (Fig. 1-3). The cytoplasm between organelles is called the hyaloplasm, and is generally a watery medium that contains ions and molecules in solution and in suspension. It serves as a reservoir of nutrients. Among the suspended molecules are protein filaments, which are responsible for the sol-gel (solution-gelatin) structure of the cytoplasm (Fig. 1-3a). Gelatin is a fine network of microscopic

oxidized, or changed to other substances. Structures change shape and location and are even destroyed and formed again in that phenomenon known as life.

Endoplasmic Reticulum

The endoplasmic reticulum (Figs. 1-1, 1-4) is a network of membranes which form tubes that pass through the cytoplasm. These membranes are continuous with the cell membrane, the nuclear membrane, and possibly the Golgi apparatus, and are believed to provide rapid transport of substances through the cell. Although these membranes at first were thought to exist only in the central cytoplasm or endoplasm, (hence the name endoplasmic reticulum), they also occur in the peripheral cytoplasm or ectoplasm. The passages

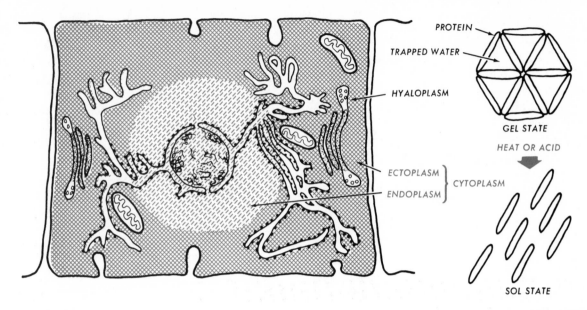

Fig. 1-3. Parts of the cytoplasm.

Fig. 1-3a. Protein fibers.

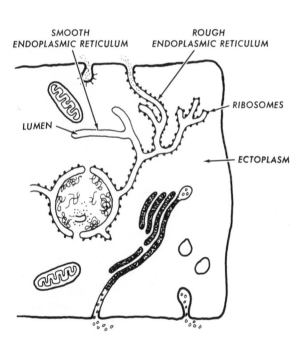

Fig. 1-4. Types of endoplasmic reticulum.

are formed by lipoprotein membranes similar to the cell membrane, and form channels which divide the cytoplasm into two phases. These are (1) the passages or lumen of the canals and (2) the hyaloplasm, which is the cytoplasm outside the membranes. The endoplasmic reticulum increases the absorption surface of the cell membrane and, like the cell membrane, can restrict the passage of certain molecules.

There are two types of endoplasmic reticulum: rough and smooth. Both types are present in most cells, but the proportion varies among the cell types. Rough endoplasmic reticulum contains small granules called ribosomes on the cytoplasmic side of the membranes. Ribosomes are involved in protein synthesis. It is believed that this synthesis can occur more easily near the endoplasmic reticulum because the channels carry solutions of the necessary raw materials. It is therefore logical that the rough endoplasmic reticulum is well developed in cells which secrete proteins such as mucus. Smooth endoplasmic reticulum contains no ribosomes, but enzymes are attached to the surface. The smooth endoplasmic reticulum predominates in acid secreting gastric cells, in steroid hormone producing cells of the testis and corpus luteum, and in liver cells involved in glycogen production and storage.

Mitochondria

A mitochondrion is a structure that is about the size of a bacillus, a type of bacteria. It is sometimes called the powerhouse of the cell. Mitochondria are numerous and produce about 95% of the cell's energy. Their number varies with the type of cell. As shown in Fig. 1-1, the mitochondrion usually is rod-shaped, and is formed by an inner and an outer lipoprotein membrane similar to the cell membrane. The inner membrane forms the inner crests, which increase the surface area of the membrane. The mitochondrial membrane, like the cell membrane, is selectively permeable. As shown in the insert in Fig. 1-5, the crests have

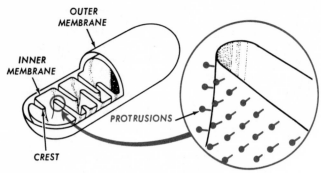

Fig. 1-5. Mitochondrion.

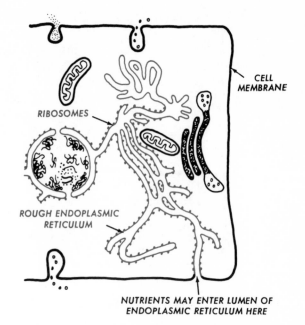

Fig. 1-6. Close-up of rough endoplasmic reticulum.

knobbed stalks, believed to be enzyme complexes, which protrude from their surfaces. The mitochondrial enzymes serve primarily in the oxidation of nutrients and in the production of energy. Mitochondria are similar to the whole cells because they also contain genetic material in the form of DNA. In fact, a mitochondrion can reproduce because this strand of DNA carries the genetic code for a new mitochondrion. This phenomenon does not take place during cell division, however, but after the two daughter cells are separated. During mitosis, the mitochondria are randomly segregated into each half of the cell.

Because of their remarkable similarities to bacteria, namely, their bacillus-like size, their DNA strand, and their ability to produce energy, some biologists speculate that the mitochondria might have originated as parasitic bacteria which invaded larger cells. As time passed, the two evolved so that they came to depend on each other for survival.

Ribosomes

Ribosomes are tiny spheroidal organelles which occur in large numbers in all cells. Their name is derived from the suffix *-some,* meaning a small body, and the prefix *ribo* from the term ribonucleic acid or RNA. Ribosomes contain a high proportion of RNA, and they are actively involved in protein synthesis. These structures are capable of decoding the genetic message contained in messenger RNA and manufacture proteins according to those specifications.

In human cells, ribosomes are usually attached to membrane structures found in the cytoplasm. They are located on the cytoplasmic surface of the endoplasmic reticulum, and constitute the rough endoplasmic reticulum (Fig. 1-6). Therefore they have great accessibility to nutrients and raw materials needed for protein synthesis. However,

in tumor or embryonic cells, these ribosomes may lie free in the cytoplasm.

Lysosomes

Lysosomes are structures about the same size and shape as mitochondria, but they have no crests. They are bounded by a lipoprotein membrane (Fig. 1-7) and contain powerful digestive juices. The term lysosome is derived from the word *lysis,* which means to break up, and the

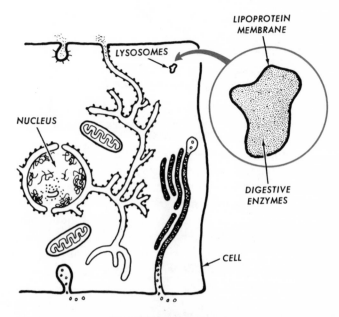

Fig. 1-7. Lysosome.

suffix -*some,* meaning small body. This is precisely the function of a lysosome. It breaks up molecules and even cells with its powerful digestive juices. These structures act as primitive digestive systems which digest the particles of cellular debris when they enter the membrane (Fig. 1-8). It is also possible for lysosomes to secrete small amounts of these digestive enzymes into the cytoplasm. This process serves to destroy cellular debris (Fig. 1-9).

The lysosomal membrane serves as a barrier to prevent the enzymes from destroying the cell itself. In a cell that is injured or dying, the sac-like lysosomes rupture spontaneously, lysis or breakdown occurs, and the cell actually digests itself (Fig. 1-10). This process, called autolysis, efficiently removes useless structures. Certain physical conditions, such as freezing and thawing or the presence of detergents, can cause autolysis. Lysosomes are present in all cells, especially in those

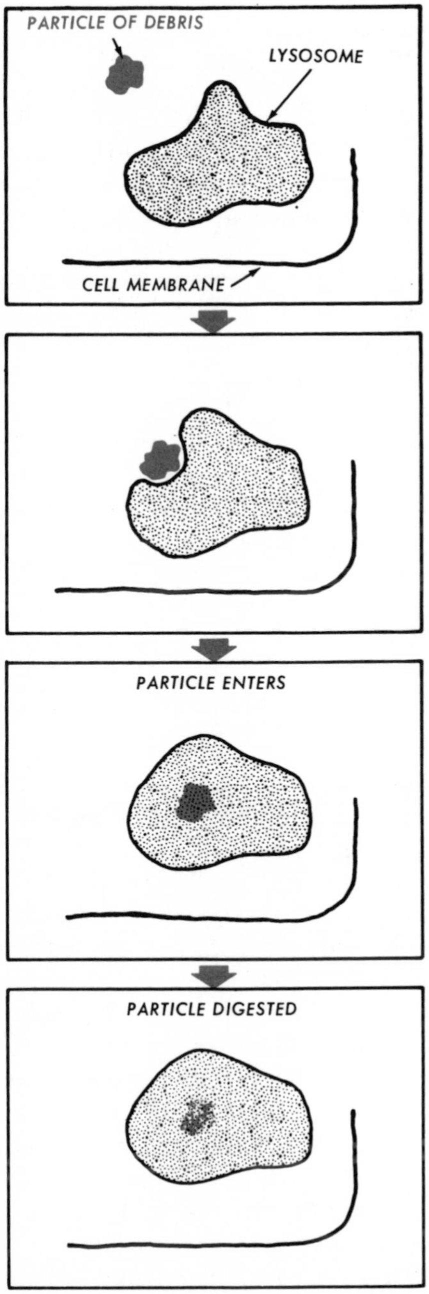

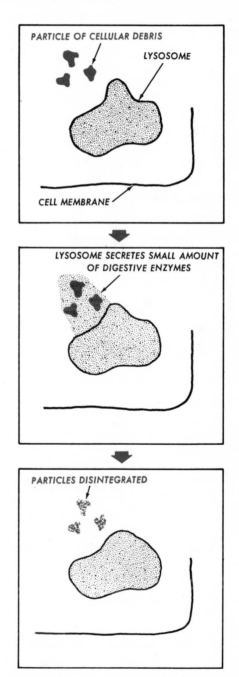

Fig. 1-8. Digestion of a small particle by a lysosome.

Fig. 1-9. Digestion of larger particles by a lysosome.

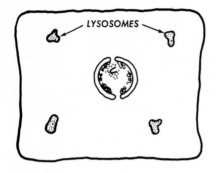

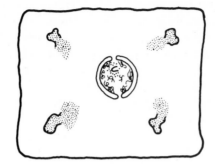

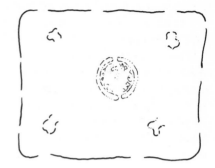

(A) Old cell or cell injured by freez-
ing-thawing or detergents.

(B) Rupture of lysosomes, release of
digestive enzymes.

(C) Autolysis.

Fig. 1-10. Autolysis of a dying cell.

cells whose purpose it is to destroy other cells. The defense mechanism of white blood cells is an example. A white cell engulfs its prey, which then enters the cytoplasm and is broken down by digestive juices released by the lysosome. However, the cell wall of certain dangerous bacteria is resistant to the digestive process and the white cells are of little defense value.

Golgi Apparatus

The Golgi apparatus was named for its discoverer, who first described it as a darkly stained mass near the nucleus of the cell. Electron microscope studies reveal that the Golgi apparatus is actually a series of parallel convolutions of a membrane similar to the cell membrane. It is of double thickness, like the endoplasmic reticulum, and is continuous with it (Fig. 1-11). The Golgi

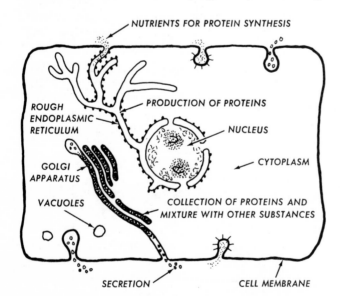

Fig. 1-11. Location of the Golgi apparatus.

apparatus is usually located near the nucleus and its surface does not have ribosomes attached. This structure is present in all animal cells, but it is most developed in the secreting cells. It is believed that the Golgi apparatus is more involved in collecting secretion products than in actually manufacturing them. For example, when the secreted product is a combination of a protein and a sugar, the Golgi apparatus concentrates the protein produced by the ribosomes and mixes it with the sugar produced in the cytoplasm and secretes the new compound.

Some investigators believe that because of their similarity the membranes of the cell itself —the endoplasmic reticulum, the Golgi apparatus, and the nucleus—are all derived from the cell membrane.

Cytoplasmic Inclusions

As we have seen, the cytoplasm contains the organelles and the clear hyaloplasm in between. There are also nonliving structures, known as cytoplasmic inclusions, that make up the cytoplasm. These cytoplasmic inclusions are the accumulated products of the various metabolic processes of the cell. They are different from organelles in that they are not living and have no active function. There are different types of inclusions: secretions, accumulations, excretory products, and storage products.

Secretions are substances synthesized and temporarily retained in the cytoplasm until released from the cell. Sometimes they are bounded by membranes, but preparation techniques make them appear as granules. Accumulations are produced by the cell for internal use, and include pigments and contractile fibers, each specific for the type of cell and its function. Storage products can be in the form of nutrient reservoirs, granules, crystals, or droplets bounded by membranes.

When a droplet is encased in a membrane, it is called a vacuole (Fig. 1-11). A lysosome is an example of a vacuole. Vacuoles are found around the Golgi apparatus and the endoplasmic reticulum. Cellular food supplies are stored in vacuoles and, during periods of starvation, human beings and other animals can live on this stored food. The fat vacuoles in certain cells of the body are examples of this storage mechanism. Sometimes fat vacuoles are so large that they occupy most of the cell. When a fat-storage cell is examined under a light microscope, only the outer membrane is visible because the preparation technique dissolves the fat and leaves a clear space. This is characteristic of fatty or adipose cells.

The Centrioles

The centrioles are a pair of short cylinders in the cell. They are essential to cell division. For a description of these organelles and the role they play in cell division, see Page 33.

The Nucleus

The nucleus was first discovered in the early 1830's by Robert Brown and scientists quickly learned that the cytoplasm needs a nucleus to survive and vice versa. The nucleus is the seat of heredity and the control center of cellular activity. It is usually located near the center of the cell (see Fig. 1-1), although in certain cell types it may be closer to the cell membrane. The nucleus is contained by the nuclear membrane or nuclear envelope which is made of liproproteins (Fig. 1-12). This membrane is continuous with the endoplasmic reticulum and, like the endoplasmic reticulum, it can have ribosomes attached to its outer surfaces. The membrane is double and the space between the membranes is filled with fluid. The nucleus is not completely sealed off by this envelope; there are pores so large that molecules, such as RNA, can pass between the nucleus and the cytoplasm.

The nucleus contains chromosomes and the chromosomes contain genes. These chromosomes

are visible with the light microscope during cell division.

The nucleus contains at least one, and usually two, nucleoli (singular, nucleolus), which appear after each new cell division. These structures are produced by special genes and are the site of RNA synthesis. The nucleoli are composed of superfine granules and filaments, but their chemical composition varies so that all cells do not stain in the same way. The nucleoli are found in the nuclear sap. They are more easily visible in some cells than in others.

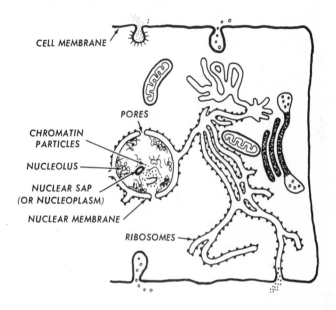

CELL MEMBRANE

PORES

CHROMATIN PARTICLES

NUCLEOLUS

NUCLEAR SAP (OR NUCLEOPLASM)

NUCLEAR MEMBRANE

RIBOSOMES

Fig. 1-12. Nucleus between cell divisions.

The nuclear sap or nucleoplasm (also called karyoplasm) is an aqueous medium, probably continuous with the hyaloplasm. It contains ions and nutrients in solution or in suspension. The electron microscope shows that the nucleoplasm contains dispersed particles and filaments. The nucleoplasm, which is in the gel state when the cell is not dividing, changes to the sol form when division occurs in order to facilitate the movement of molecules.

Chromatin particles are the undissolved parts of chromosomes seen when the cell is not dividing. Chromosomes are visible only when the cell is getting ready to divide. At other times they "unravel" and are dispersed in the nucleoplasm. The chromatin bodies are the parts of the chromosomes which did not dissolve.

Review Questions

1. Label the lipid layer(s), and the protein layer(s) in the illustration below:

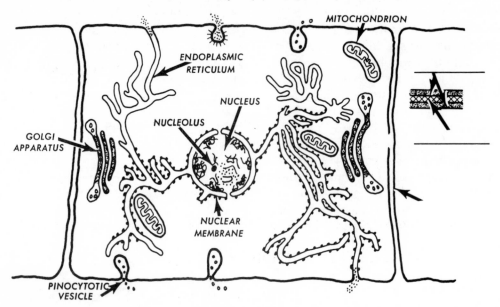

2. Which of the following statements are true about selective permeability?
   a. Some substances can enter freely, others cannot.
   b. The membrane acts as a filter—all large molecules are held back, all small molecules **may enter** freely.
   c. It depends on the molecule's ability to dissolve in lipids.
   d. It depends on the electrical charge of the molecule.

3. What word is used to describe small, finger-like projections of the cell membrane? _____

_____

4. Define desmosome: _____

5. Label the hyaloplasm, the endoplasm, the ectoplasm in the illustration below:

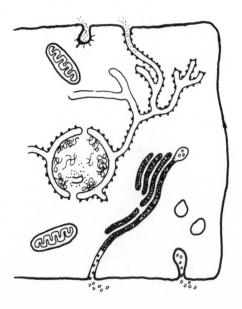

6. Label the parts in the illustration shown below:

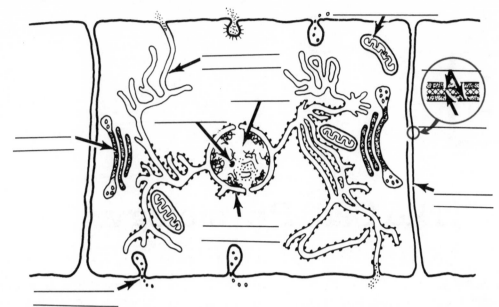

7. Name the granules on the rough endoplasmic reticulum. _____
_____

8. What is the powerhouse of the cell? _____

9. What is the main function of mitochondrial enzymes? _____
_____

10. The main function of ribosomes is _____
_____

11. What intracellular structures contain such powerful digestive juices that they can break up cellular
debris ? _____

12. Give another word for this type of "breaking-up." _____

13. The Golgi apparatus is involved in:

    a. Autolysis   c. Collection of secretion products

    b. Energy production

14. A droplet encased in a membrane is known as a _____

15. Label the parts in the illustration shown below:

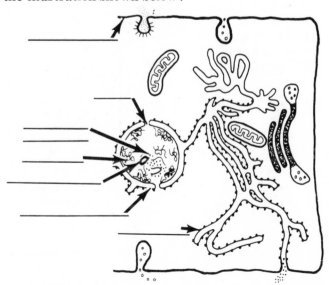

chapter 2

The Cell: Physiology

In order for a cell to survive, it must be able to acquire nutrients and remove wastes. In the human body, the blood supplies food and oxygen and removes wastes. How does this exchange take place at the cellular level? We learned earlier that the cell membrane is selectively permeable so that not all substances can pass across it. We also learned that the membrane is a living part of the cell and controls the amount of a substance that enters or leaves the cell. We shall approach this problem of exchange in two ways: first by discussing the question of crossing the membrane itself, then by learning about another transport system using vacuoles.

Membrane Transport

The cell membrane controls the internal environment of the cell, while the other membranes inside the cell control the individual environments of the organelles, such as the mitochondria, nucleus, ribosomes, lysosomes, Golgi apparatus, and endoplasmic reticulum. So it is reasonable that each of these organelles has its own individual environment concerning acidity, concentration of water, or other substances. We will use the cell membrane as a model, although the other membranes probably utilize similar control mechanisms.

A generation ago it was believed that the cell membrane was a simple sieve which strained out large molecules to keep them from entering the cell. Today, we know that the mechanism is much more complex. At this point we should discuss the differences between diffusion and osmosis. Diffusion is the tendency of molecules to disperse themselves throughout a given space. In chemistry you learned that a gas will expand to fill its container. For our purposes, we will consider an aqueous medium, that is, water. All molecules have a certain random motion; they vibrate in all directions. The agitation speeds up as the temperature increases. When many of these molecules are close together they collide and push away from each other. Fig. 2-1 shows how a sugar solution added to a container of water will disperse itself equally throughout the water, through strictly random movement, known as diffusion. If we heat the water, the diffusion occurs more quickly because the molecules move more rapidly and therefore diffuse more quickly. At the same

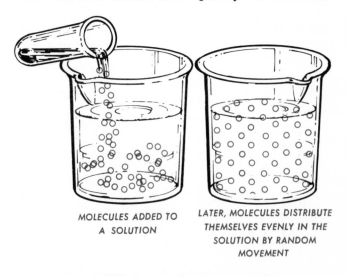

MOLECULES ADDED TO A SOLUTION

LATER, MOLECULES DISTRIBUTE THEMSELVES EVENLY IN THE SOLUTION BY RANDOM MOVEMENT

Fig. 2-1. Diffusion of molecules in a solution.

time, the water molecules also are vibrating and diffusing through the other substance.

The cell membrane is built so that water can freely diffuse across it in either direction. Gases such as carbon dioxide and oxygen can also freely diffuse across the cell membrane. There are other substances, however, which can freely diffuse in water, but cannot pass across this membrane barrier. Probably one of the reasons for this limitation is the electrical charge.

There are two ways to cross the cell membrane: passive transport and active transport. The main difference between these two mechanisms lies in the fact that passive transport is simple diffusion and requires no effort on the part of the cell, but active transport requires the expenditure of a certain amount of energy.

The most common form of passive transport is osmosis, which is defined as the passage of water across a membrane. We use the term diffusion when we speak of substances other than water crossing the cell membrane.

Since water molecules can pass freely across the membrane, but other molecules cannot, we can imagine an imbalance on each side of the membrane. Fig. 2-2a, b, c shows a membrane barrier with water on either side. Both sides have the same concentration of large molecules in solution. When large molecules are added to the right side of the barrier they cannot pass to the left, but the water can go either way and begins to migrate to the side where the molecules are, through random movement. This phenomenon is called osmosis. Fig. 2-3 is a more realistic example. Fig.

2-3a shows a red blood cell in a salt solution. The water concentration in the solution is equal to that of the cell, so we call this situation isoosmolarity because the osmotic properties are equal on both sides of the cell membrane. Fig. 2-3b shows the same situation, but water is added, diluting the solution. The solution becomes hypoosmotic or a hypotonic solution because there is a greater proportion of water outside the cell. Since the molecules inside the cell cannot pass out through the membrane, the water outside must pass into the cell to reach equilibrium, causing the cell to swell and burst. Fig. 2-3c shows the results of adding salt to the solution. This decreases the water concentration, producing a hyperosmotic or hypertonic solution and causes the water to leave the cell. The cell shrinks and eventually dies of dehydration. Remember it is the water concentration, not the substances, that is important. Different substances in the same concentrations on either side of the membrane can still reach equilibrium.

Active Transport

The living cell must acquire nutrients, expel wastes and maintain its steady state; activities which sometimes require energy. This mechanism is known as active transport because it actively moves molecules against the normal direction of osmotic movement, "upstream" so to speak, or even helps molecules to go "downstream" in the normal direction, only faster. These active transport mechanisms are known as "pumps," and are

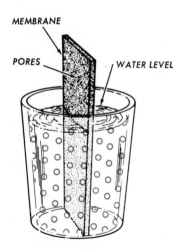  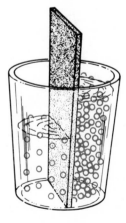

(A) Equilibrium.   *(B) Addition of nondiffusible molecules on one side.*   *(C) Water migrates to equalize the concentrations.*

Fig. 2-2. Simple osmosis.

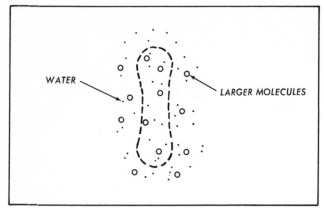

(A) Red blood cell in salt solution.

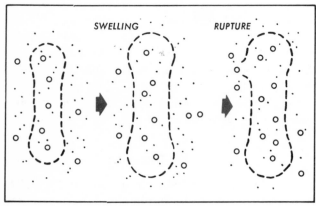

(B) Hypotonic solution.

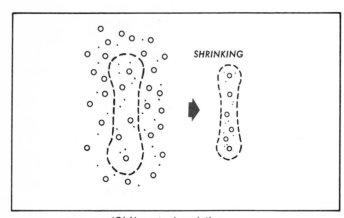

(C) Hypertonic solution.

Fig. 2-3. Osmotic process.

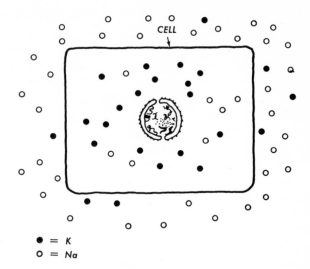

● = K
○ = Na

Fig. 2-4. Relative concentrations of Na and K inside and outside the cell.

specific to certain substances. Sodium, for example, is moved by the sodium pump. In mammals, both sodium (Na) and potassium (K) are present in cells and in the fluids around the cells. Inside the cell, potassium is present in a greater concentration than sodium, while the opposite is true in the fluids, such as blood, surrounding the cell (Fig. 2-4). The cell must continually and actively remove sodium from the cell and bring in potassium so that the proper levels can be maintained. These differences between cell fluids and extracellular fluids are essential to cell life processes. When certain poisons block energy production in the cell the active transport stops, the cells lose their equilibrium, and die.

Vacuole Transport

In this type of transport, the substances to be transported are "packaged" in a vacuolar membrane. There are three types of vacuole transport: pinocytosis, the ingestion of liquid material; phagocytosis, the ingestion of particles; and exocytosis, the removal of substances, especially in secretion.

First we will discuss the two mechanisms involved in ingestion or intake of substances. There are certain points on the surface of the cell membrane called binding sites, where molecules can attach themselves (Fig. 2-5a). When a molecule attaches to one of these sites, it causes the membrane to react by invaginating or cupping inward (Fig. 2-5b). As the membrane invaginates, it surrounds the molecule and finally isolates it completely as a vacuole bounded by a membrane (Figs. 2-5c, 2-5d). If the molecules ingested are liquid, the process is called pinocytosis (from Greek words that mean drinking). If the substance is a solid particle, the process is called phagocytosis (from Greek words that mean eating). The solid particles enclosed in this vacuole then are digested by the cell, usually by the lysosomes.

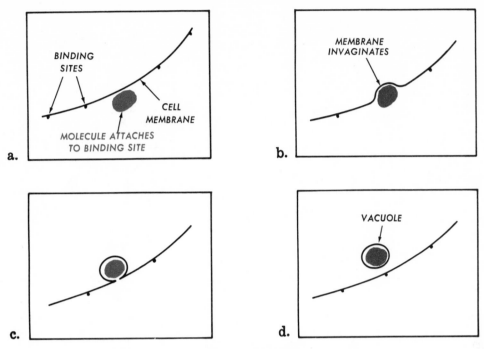

Fig. 2-5. Pinocytosis.

Exocytosis, or removal to the outside, is the process whereby substances or molecules too large for either diffusion or active transport are transported out of the cell. It is the opposite of pinocytosis and phagocytosis. The process of secretion is the most common example of this phenomenon.

Review Questions

1. Indicate true statements with "T" and false statements with "F"
   a. _____ Water molecules will diffuse to fill any size space.
   b. _____ A form of active transport is osmosis.
   c. _____ Osmosis is the passage of particles across a membrane.
   d. _____ Diffusion is the passage of particles across a membrane.

2. Match the words with the appropriate phenomena in the illustration below:
   a. Hypertonic
   b. Hypotonic
   c. Hypoosmotic
   d. Hyperosmotic
   e. Isotonic

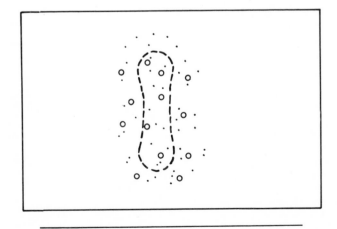

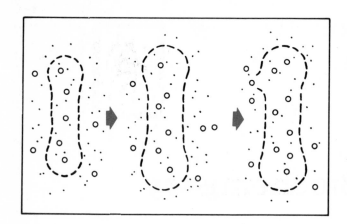

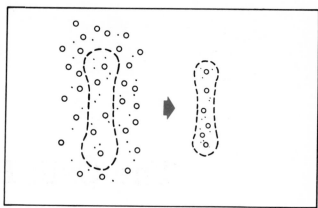

_____   _____

3. Insert these labels in the illustration below.
   a. Sodium molecules   b. Potassium molecules

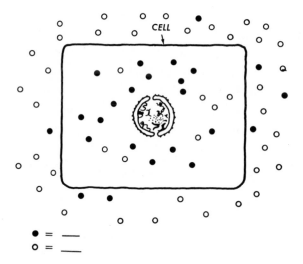

● = ____
○ = ____

4. Match the following:

   Pinocytosis   _____   a. taking in of liquids
   Exocytosis   _____   b. taking in of solids
   Ingestion   _____   c. removal of substances
   Phagocytosis _____

chapter 3

The Cell: Biochemistry

You may recall that when you studied chemistry there were many times when you had to heat substances in a test tube for reactions to occur. The reason was that a certain minimum amount of heat—known as the energy of activation—was required to start the reaction. You may have used catalysts, such as fine metal powders, in some of your experiments. A catalyst is a substance that increases the rate of a reaction so the result becomes apparent. Usually, you would use a catalyst to speed up a reaction which would have occurred anyway, but at a rate so slow that years would pass before the results were apparent. A catalyst becomes involved with the reaction, but it is never changed by the reaction. When the reaction is over, you can collect the catalyst and find it unaltered and able to be reused. If two substances which cannot react together are mixed with a catalyst, the reaction will not occur, because a catalyst cannot make a reaction occur. It merely speeds it up so the results are evident.

Enzymes

In biology, living catalysts are called enzymes. Enzymes are composed primarily of proteins and act in exactly the same manner as the catalysts in the chemistry experiments. Their mechanism of action is simple; they help reactions to occur at body temperature, whereas such reactions normally would require more heat. It would not be practical to heat a cell over a Bunsen burner each time a reaction was to occur, so enzymes make it possible for reactions to take place at body temperature. The substance the enzyme acts upon is known as its substrate and the result of the reaction is known as the product. In most cellular reactions, the enzyme is specific to a given substrate and is named accordingly, with the suffix -ase which denotes an enzyme. For example, the enzyme which catalyses the reaction that makes possible the removal of hydrogen from glucose-6-phosphate is known as glucose-6-phosphate dehydrogenase. Sometimes the enzyme is named after the type of reaction or the product. The important thing to remember is that a cell is the site of hundreds of different chemical reactions, each of which is catalysed by an enzyme.

Energy

The living cell needs energy for membrane transport, for movement, and for chemical synthesis. Our energy is derived from foods which are broken down. When a nutrient is degraded, energy is released and captured in chemical bonds, mainly phosphate compounds. Since it takes energy to attach a phosphate to a molecule, this energy is stored in the chemical bond and then released when the bond is broken. An exact amount of energy is needed to form each bond, and the energy left over is released as heat, a relatively useless form of energy for the cell. Enzymes control the formation and destruction of these phosphate bonds. When energy is needed for a particular use, the appropriate enzyme breaks the bond and releases the energy; that is, it transfers the energy to another chemical bond. The only time heat is released is when some energy is left over. As you can see, this is a relatively efficient system, with a minimum of energy loss. Therefore, energy is stored in parcels in the form of chemical bonds. The most important of these compounds is known as ATP or adenosine triphosphate. ATP is the principal source of energy in the body. The mitochondria, the powerhouses of the cell, oxidize or burn the food we eat, but relatively little heat is given off because it would be a waste of energy and because it obviously would be harmful to the cell. When ATP releases its energy for a given reason, one of the phosphates is removed, energy is released, and the compound becomes ADP or adenosine diphosphate. The ADP in the mitochondria then pick up more energy and become ATP again.

Although our diet consists of a combination of proteins, fats, and carbohydrates, our main energy source is carbohydrates. Carbohydrates, through various digestive processes, are eventually broken down into glucose, a simple sugar. Fig. 3-1 shows the overall plan of glucose metabolism. The first main process the glucose undergoes is phosphorylation, where an ATP gives a phosphate to the glucose, producing glucose-6-phosphate. The glucose-6-phosphate can then take one of two paths; the choice is controlled by the two different types of enzyme systems. In the mitochondria, glucose-6-phosphate undergoes the process of gycolysis, where it is broken down into pyruvic acid. The pyruvic acid is then utilized by the Krebs cycle (also called the tricarboxylic acid cycle or citric acid cycle) where the pyruvate is broken down into H_2O (water) and CO_2 (carbon dioxide) along with energy bound to a group of compounds carrying the energy produced in the form of electrons. The energy contained in these electron arrangements is then transferred to many ADP molecules which, in turn, are converted to ATP molecules, the final product. The system which involves this electron transfer process is called the respiratory chain. The respiratory chain occurs in the mitochondria and is a major source of energy. (Although both glycolysis and the Krebs cycle produce ATP molecules, most are derived from the respiratory chain.)

Another pathway, called the pentose phosphate pathway, exists in the cytoplasm and this too con-

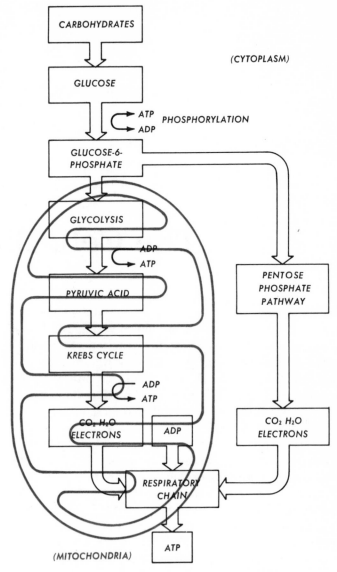

Fig. 3-1. Overall plan of glucose metabolism.

verts the glucose-6-phosphate to other compounds, finally ending up with H_2O, CO_2, and electron transfers which also pass through a respiratory chain to produce ATP from ADP molecules.

These two processes are the means by which an aerobic cell (one which requires oxygen) gets its energy. All the cells of the body use these pathways for energy, but some tumor cells are anaerobic and, using the glycolysis pathway, break the glucose-6-phosphate into pyruvate, which is then converted to lactate.

Review Questions

1. Match the following:
   a. Substrate
   b. Product
   c. Catalyst
   d. Energy of activation

   _____ Substance that increases the rate of a reaction so it becomes apparent

   _____ the substance an enzyme acts on

   _____ minimum amount of heat required to start a reaction

   _____ result of a reaction

2. In the cell, energy is principally stored as : _____
   a. ADP
   b. ATP
   c. Neither
   d. Both

3. Glucose passes through several steps to produce ATP. Place the following steps in the proper sequence:

   Krebs cycle _____   Pyruvic acid _____

   Glucose _____   Glycolysis _____

   Respiratory chain _____   ATP _____

   Glucose-6-phosphate _____

4. Match the following:
   a. Aerobic metabolism
   b. Anaerobic metabolism

   _____ requires oxygen

   _____ less efficient, requires more food to produce the same amount of energy

   _____ used by some tumor cells

chapter 4

The Cell: Genetics

Every process that occurs in the living cell is controlled by substances called nucleoproteins, which are found in the cell's nucleus. These chemical strands regulate all the cell's activities by controlling the synthesis of enzymes and by regulating the activity of these enzymes. The nucleoproteins are the patterns by which the cell is controlled. The word nucleoprotein obviously is derived from the location of these chemical strands in the cell and the fact that they are composed of protein. You have no doubt read about the nucleoproteins DNA (deoxyribonucleic acid) and RNA (ribonucleic acid). They are believed to have first appeared millions of years ago in the first bits of matter regarded as living. Nucleoproteins are chemicals that can reproduce themselves. They control the synthesis of proteins. Proteins make up enzymes, the structural framework of the cell hormones, and the antibodies. Proteins control their own synthesis, under the direction of DNA, through enzymes.

Nucleic Acids

A protein is a chain of nitrogenous substances called amino acids. A nucleic acid is more complex; it is a chain of nucleotide units. Each unit, or nucleotide, consists of three components: a pentose sugar, a nitrogenous base, and a phosphate group. There are several differences between DNA and RNA. First of all, DNA issues the "orders" and RNA transmits them to the cell. The pentose sugar in DNA is called deoxyribose, hence the name deoxyribonucleic acid. RNA has the pentose sugar ribose in its structure. The nucleotides also differ between the DNA and RNA, each one being made up of four groups. DNA nucleotides are adenine, guanine, cytosine, and thymine, abbreviated as A, G, C, and T. RNA contains the nucleotides adenine, guanine, cytosine, and uracil (A, G, C, U). Therefore, they have three of the nucleotide groups in common, the only difference being the thymine in DNA and the uracil in RNA. We will first discuss the structure and role of DNA and then show the relationship of DNA to RNA.

DNA

The DNA molecule is the memory bank of the cell. It consists of a double strand of nucleotides arranged in a code, called the genetic code. There are infinite combinations of the four nucleotides, which are strung along the length of the DNA molecule. Fig. 4-1 shows that the DNA molecule consists of two strands, lying exactly parallel to each other, both in the form of a spiral. The two chains are linked together by horizontal bonds, like the rungs of a ladder, matching the nucleotides to one another across the two strands. You will notice that the A always is attached to the T nucleotides and the C is always attached to the G nucleotides. The nucleotides are built so that they can only attach to their complements, A-T, C-G. This phenomenon is the basis of the genetic code. Looking at the DNA molecule represented in the figure, you note that the two strands are not identical, but complementary; one is the answer

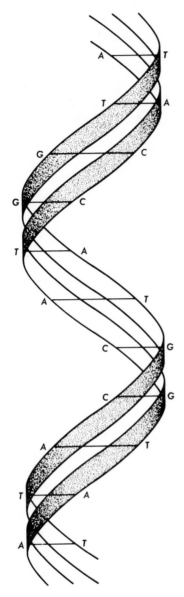

Fig. 4-1. DNA molecule.

As you recall from the discussion of the nucleus in Chapter 1, the chromosomes are visible only during cell division. Each chromosome is a length of DNA, and before the cell begins to divide, the DNA molecule divides. This occurs while the chromosomes are still unravelled in the nucleoplasm. DNA must duplicate itself exactly in order for any meaningful genetic message to be inherited by the dividing cells. A mistake which occurs during the DNA duplication is known as a mutation, and the cell usually does not survive. It is rare for a mutation to survive and even rarer for a mutation to be a biological improvement, although this has been known to happen.

Briefly, the DNA duplication involves an unzipping of the double chain and the copying of each of the chains. Fig. 4-2 shows how the molecule unzips at one end and how each half takes on new

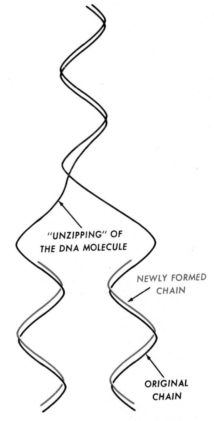

"UNZIPPING" OF THE DNA MOLECULE

NEWLY FORMED CHAIN

ORIGINAL CHAIN

Fig. 4-2. DNA "unzipping" and replication.

to the other. The nucleotides can appear on either side and in any order, but they must always be linked up A-T, C-G.

It has been proven experimentally that a given segment of the DNA molecule is responsible for a given protein, and the combination of nucleotides forms the code by which that particular protein is synthesized. It has been learned that each three nucleotides, or triplet, such as AGA, constitutes a letter of the code, and that still other triplets are the equivalent of the word "stop" in a telegram and halt the reading of the code. Before discussing the genetic code further, let us look at how the DNA molecule duplicates itself.

complementary nucleotides, so that each of the daughter molecules has one original strand and one newly formed strand. All the attaching, unzipping, and bond formation is done with specific enzymes. Each enzyme is specific for a given step in the complicated process. Of course, energy is required to form these new bonds.

The Genetic Code

The Morse code has only two symbols, a dot and a dash, yet every word of our language can be transmitted with it. The genetic code has four symbols—the nucleotides. Each combination of three nucleotides is equivalent to a letter. G-C-A, for example, is the code for the amino acid alanine. Since a protein is a chain or polymer of amino acids, each combination of three nucleotides stands for an amino acid and each length of the DNA molecule, known as a gene, carries the code for a certain protein (Fig. 4-3).

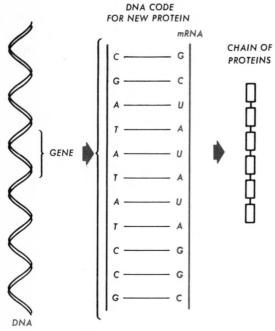

Fig. 4-3. Decoding a gene.

Messenger RNA, abbreviated mRNA, acts as the intermediary between the code on the DNA molecule and the cytoplasm of the cell. You recall that proteins are synthesized in the cytoplasm by ribosomes. Fig. 4-4 shows how mRNA is produced. When a certain protein, an enzyme for example, is needed by the cell, the part of the DNA molecule carrying the "recipe" for the enzyme partially unravels. At this point, nucleotides are added to a strand in a way similar to the process of DNA replication, except the nucleotides are paired up as C-G (as in DNA synthesis) and A-U (remember, RNA contains uracil not thymine). Thus, a strand is formed from the temporarily separated DNA strand and then is passed from the nucleus to the cytoplasm. Ribosomes, which are the sites of protein synthesis, have to "read"

the recipe written on the mRNA so they can make the protein needed by the cell at that moment. Each ribosome attaches at the end of the mRNA and then travels across the strand, reading the message and making the required protein. As each ribosome completes its trip along the mRNA molecule, it separates from the strand and releases its protein product. Several ribosomes can follow one another across the strand, as shown in Fig. 4-5.

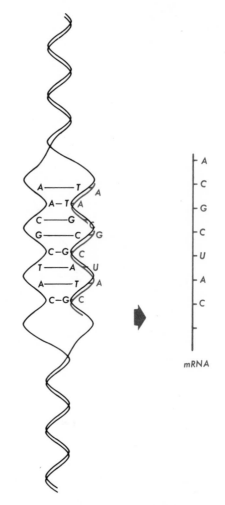

Fig. 4-4. mRNA production.

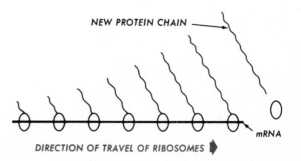

Fig. 4-5. Protein synthesis.

The amino acid building blocks are present in the cytoplasm, especially near the endoplasmic reticulum, as we mentioned earlier. They can diffuse into the cytoplasm and into the ribosomes themselves. How are they brought to the proper place on the new protein chain? Exactly how does the ribosome read the message on the mRNA? Another type of RNA, called transfer RNA or tRNA, is responsible for these phenomena. tRNA is a very short RNA molecule which attaches itself to amino acids. Each type of tRNA is specific for a certain type of amino acid (there are 20 types). Imagine the tRNA molecule as having one end specific to a particular amino acid and the other end specific to a given triplet. Fig. 4-6 shows the process in which the tRNA attaches itself to the amino acids floating in the ribosome and then delivers them to the corresponding triplet. There-

fore, the tRNA is the complement to the messenger RNA and can attach itself to the strand. Meanwhile, the amino acid it carries can be attached to the end of the chain being formed, using specific enzymes. This is the way the same message can be carried for generations, even millions of years. Fig. 4-7 summarizes DNA functions.

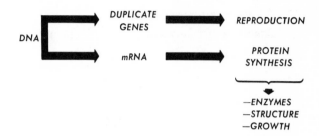

Fig. 4-7. Summary of the functions of DNA.

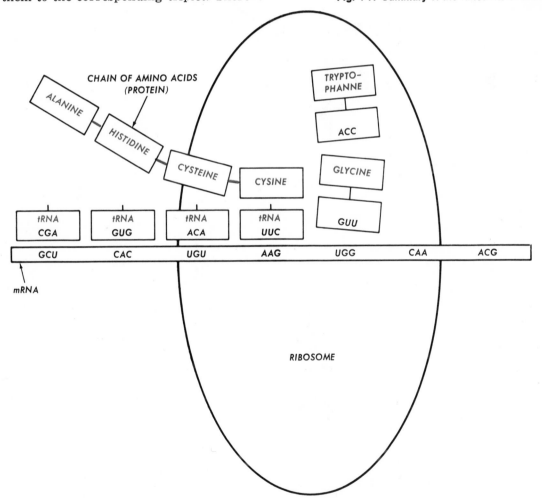

Fig. 4-6. Attachment of tRNA to amino acids in the ribosome and to the corresponding triplet.

Review Questions

1. Match the following with the substances named below: a. Nucleoprotein   b. Part of a nucleo-protein   c. Neither

   _____ Guanine   _____ Cytosine   _____ Deoxyribose   _____ Thymine

   _____ Adenosine triphosphate   _____ Adenine   _____ Ribonucleic acid   _____ Nucleotide

   _____ Deoxyribonucleic acid   _____ Ribose   _____ Uracil

2. In the illustration below:
   1. Fill in missing nucleotides
   2. The drawing shows a: _____ molecule

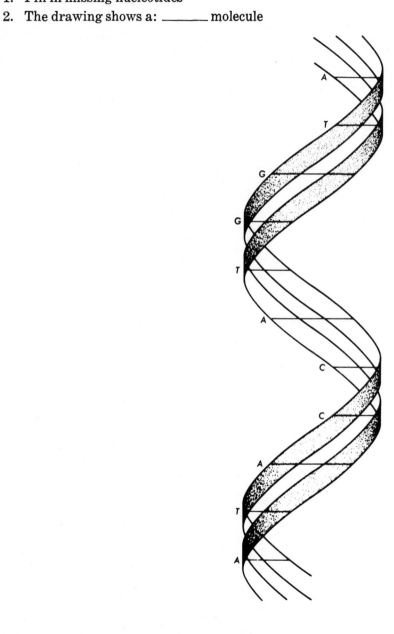

3. How would you convert the drawing to show RNA?

4. In the illustration below, label:
   a. DNA   b. mRNA   c. New protein

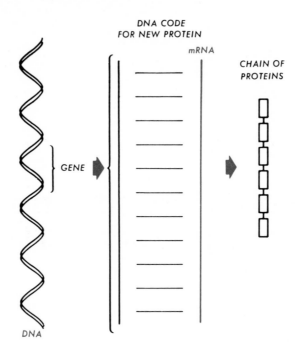

DNA CODE
FOR NEW PROTEIN

mRNA

CHAIN OF
PROTEINS

GENE

DNA

5. Match the following:
   a. DNA   _____ carries the "recipe" for a protein into the ribosome
   b. tRNA   _____ carries the amino acids for the "protein recipe" into the ribosome
   c. mRNA   _____ has ribosomes attached to it like beads on a string

chapter 5

The Cell: Reproduction and Tissue Formation

Cells are regularly being replaced by the millions. They are constantly growing, dividing, being injured, being digested, being worn away, and being crushed. Some cells live for only a few days. Others live much longer. But all cells have one characteristic in common: they need to multiply, to repair themselves, and to grow. These processes, as well as all other processes of the living cell, are controlled by the cell nucleus. The formation, structure, and function of nucleoproteins and their role in genetics were discussed in Chapter 4. The reproduction of cells by the observable phenomenon of cell division, called mitosis, and the grouping of cells into tissues is the subject of this chapter.

Chromosome Number

Each of the cells of the body contain 46 chromosomes, except for the reproductive or germ cells. A germ cell in the male is called a sperm cell and a germ cell in the female is an egg or ovum. Cells other than the reproductive cells are called somatic cells or body cells. It is these somatic cells that contain 46 chromosomes; the germ cells contain only half that number, or 23 chromosomes. When an egg is fertilized by a sperm, each of the germ cells donates 23 chromosomes, bringing the number of chromosomes in the fertilized cell to 46. This 46-chromosome cell then multiplies by mitosis, giving rise to all the cells of the body. The term haploid is used to refer to the 23 chromosome number; the term diploid to the double number, 46.

Each somatic cell contains 23 pairs of chromosomes; 22 of the chromosome pairs are called autosomes and the remaining pair are sex chromosomes. Each of the male cells contains 44 autosomes (22 pairs) and an X and a Y sex chromosome. Each of the female cells has the same autosome number, but differs from the male cells in that they have two X chromosomes (see Fig. 5-1). The sex cells are produced by meiosis or

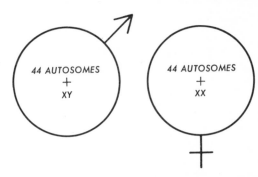

Fig. 5-1. Difference between cells in males and females.

reduction division, a process in which the normal 46 chromosome complement is divided in half for the germ cells (see Fig. 5-2). The female, therefore, can only produce the combination of 22 autosomes plus X, while the male can produce germ cells containing 22 autosomes plus X or 22 autosomes plus Y. When sperm and egg come together, the sex of the offspring is determined by which sperm, X or Y, fertilizes the egg (Fig. 5-3).

There are abnormal variations possible in the chromosome number. In one form, known as polyploidy, the cell contains an even multiple of the

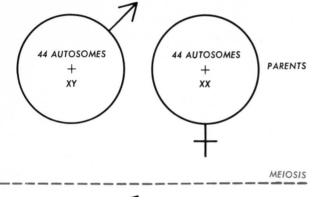

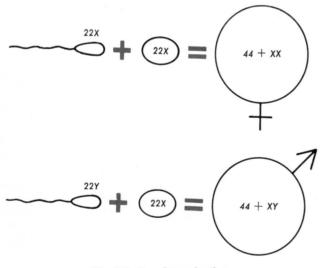

Fig. 5-2. Meiosis.

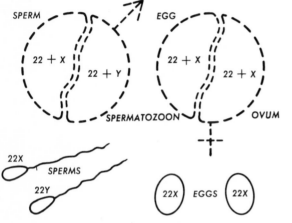

Fig. 5-3. Sex determination.

normal haploid number. Tetraploidy is an example of polyploidy, where the cell contains four times the normal haploid number of chromosomes, that is, two times the diploid number. These cells, which are characterized by a large nucleus, result from an abnormal mitosis. In aneuploidy, another variation, a somatic cell has less than the diploid

number (such as 43 chromosomes), or more than the diploid number, but not an even multiple of it. An example of the latter aneuploidy is a certain type of mongolism, which results when the somatic cells have an extra chromosome. These anomalies of chromosome number frequently involve the sex chromosomes, and sometimes can be detected by examination of chromatin bodies.

Centrioles

The centrioles are the organelles responsible for the synthesis of the spindle fibers used in mitosis. They are short cylinders, one always at right angles to the other. Before cell division can occur, they duplicate themselves and migrate to opposite poles of the cell. They play a specific role in mitosis.

Mitosis

Mitosis is the process by which somatic cells reproduce. Since diploid cells give rise to diploid offspring, or daughter cells, there must be a process by which the chromosomes can reproduce. The diagram of the cell in Chapter 1 (Fig. 1-2) represents the time in the cell life cycle called the interphase, where no apparent reproduction is taking place. In the human body, the average minimum time from one division to another division is 16 hours, although some cells reproduce much more slowly. Mitosis differs from meiosis in that it produces daughter cells with the same diploid chromosome number, while meiosis produces daughter cells with the same haploid chromosome number. Mitosis entails several steps, namely prophase, metaphase, anaphase, and telophase.

Before mitosis can occur, the chromosomes must duplicate themselves. Therefore, at the end of the interphase, or non-dividing phase, DNA duplication takes place. This part of the cell cycle is called the S (for synthesis) stage. Fig. 5-4a shows the cell at interphase. Nothing is really apparent at this point; the chromosomes are unravelled, or dispersed, in the nucleus and are surrounded by the nuclear membrane.

The prophase is shown in Fig. 5-4b. In this phase, changes occur in both the cytoplasm and the nucleus. In the cytoplasm, the two pairs of centrioles separate and travel to opposite ends of the cell. As the two pairs separate, they seem to be pulling thin filaments called spindle fibers between them. These fibers play a key role in cell division. At the same time, short fibers grow out in a starlike pattern from each pair. They are called asters (for star). These aster fibers extend from the centrioles to the peripheral part of the cytoplasm. Simultaneously other changes occur in the prophase nucleus. Fig. 5-4b shows how the nucleolus and nuclear membrane disappear. The nucleoproteins begin to condense and start to resemble chromosomes. During mitosis, chromosomes appear as short rods, because the long DNA threads coil up. The chromosomes are already duplicated and each identical pair is held together by a centromere, as shown in Fig. 5-4c. Each half of the chromosome pair is called a chromatid. Later in prophase, (Fig. 5-4d, e) each chromatid pair is attached to a spindle fiber, and begins to migrate toward the center of the spindle, or the equator of the cell. Prophase is considered complete when the chromatids reach the equator or center of the spindle.

Metaphase is the next step in mitosis, and is considered to be the midpoint of the process. Fig. 5-4f shows the spindle fully formed and the chromosomes arranged at the equator, where they

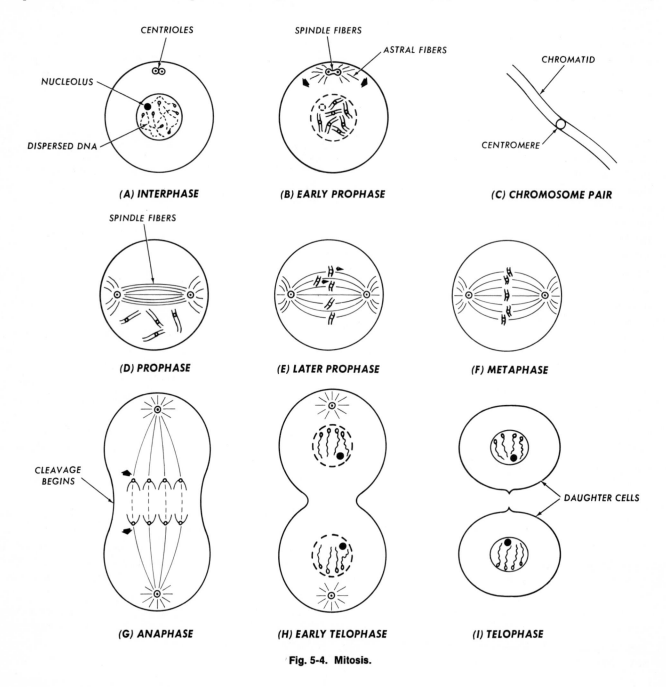

Fig. 5-4. Mitosis.

temporarily remain stationary. At this time, the centromeres divide at right angles to the spindle fibers so that a member of each chromatid pair is destined for a different half of the cell. Each chromatid is attached to a spindle fiber so that it can be pulled toward the centriole at the end of the cell. This dividing of the chromatids occurs in the late stage of metaphase.

During anaphase, each member of each chromatid pair is pulled to an opposite pole (Fig. 5-4g). This apparently is accomplished by the spindle fiber pulling the chromosome by its centromere, so that its "arms" trail behind in a V. Since there are more spindle fibers than chromosomes, the remaining spindle fibers grow longer, and the cell elongates and begins to narrow at the midpoint. The process by which the cytoplasm is divided is called cytokinesis or cleavage.

Fig. 5-4h shows telophase. In this final stage, the cytoplasm eventually pinches off into two separate daughter cells and the nucleus regains its original appearance. Fig. 5-4h shows the early part of the telophase. The cleavage becomes deeper and deeper, possibly due to the contraction of the astral fibers, which are rooted in the peripheral cytoplasm. At the same time, in the nucleus, the chromosomes begin to uncoil and the nuclear membrane and nucleolus (or nucleoli) are reformed. Some parts of the chromosomes do not uncoil; these are called chromatin granules. Fig. 5-4i shows the later stage of telophase. The cell membrane has constricted so that the two daughter cells are separate and the centrioles have reduplicated in preparation for the next division.

Sex Chromatin

As we mentioned previously, the chromosomes are uncoiled during interphase and remain invisible except for certain segments which remain tightly coiled. These are called the chromatin granules. This is most commonly referred to as female sex chromatin because it is visible in the interphase nucleus. The following explanation gives a probable reason for this phenomenon. A chromosome must unwind in order to synthesize mRNA for the expression of its genes. There are two X chromosomes in the female, but only one is needed for synthesis. The other remains tightly coiled. The male X chromosome must uncoil so it can function properly, so the sex chromatin is not visible in the male (Fig. 5-5). The presence of the sex chromatin can be interpreted to mean that

there are two X chromosomes and one is "busy." The absence of the sex chromatin can be interpreted to mean that there is only one available X, so the cell belongs to a male. However, in genetic disorders, it is possible to have several X chromosomes, such as seen in Fig. 5-6, Kleinfelter's syndrome. The patient had four X's and a Y (XXXXY), but since one of the X's was "busy," only three chromatin granules are visible.

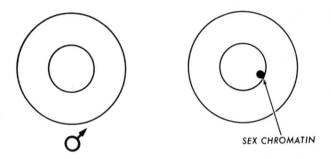

SEX CHROMATIN

Fig. 5-5. Sex chromatin in normal cells.

Fig. 5-6. Kleinfelter's syndrome.

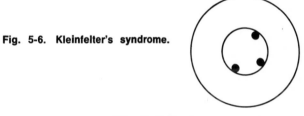

The Cell Cycle

Through repeated mitosis, cells constantly are replaced. The length of time between each mitotic division is called the cell cycle. Fig. 5-7 shows the relative time devoted to each part of the cell cycle. The S stage, when DNA replication takes place, is followed by a quiet period, called the G2 stage, or post-duplication stage. Next, mitosis occurs; and, before the S stage begins again, there is a period called the G_1 stage or presynthetic stage.

Organization into Tissues

When a fertilized egg begins to divide, it eventually forms a mass of cells with three distinctly different layers. Although all these cells still have the same genetic content, they express this content differently. These three layers of the early embryo are called the ectoderm or outer layer, endoderm or inner layer, and mesoderm or middle layer. Each layer gives rise to a different group of cells and organs. Skin, for example, is derived from the ectoderm, the digestive system comes

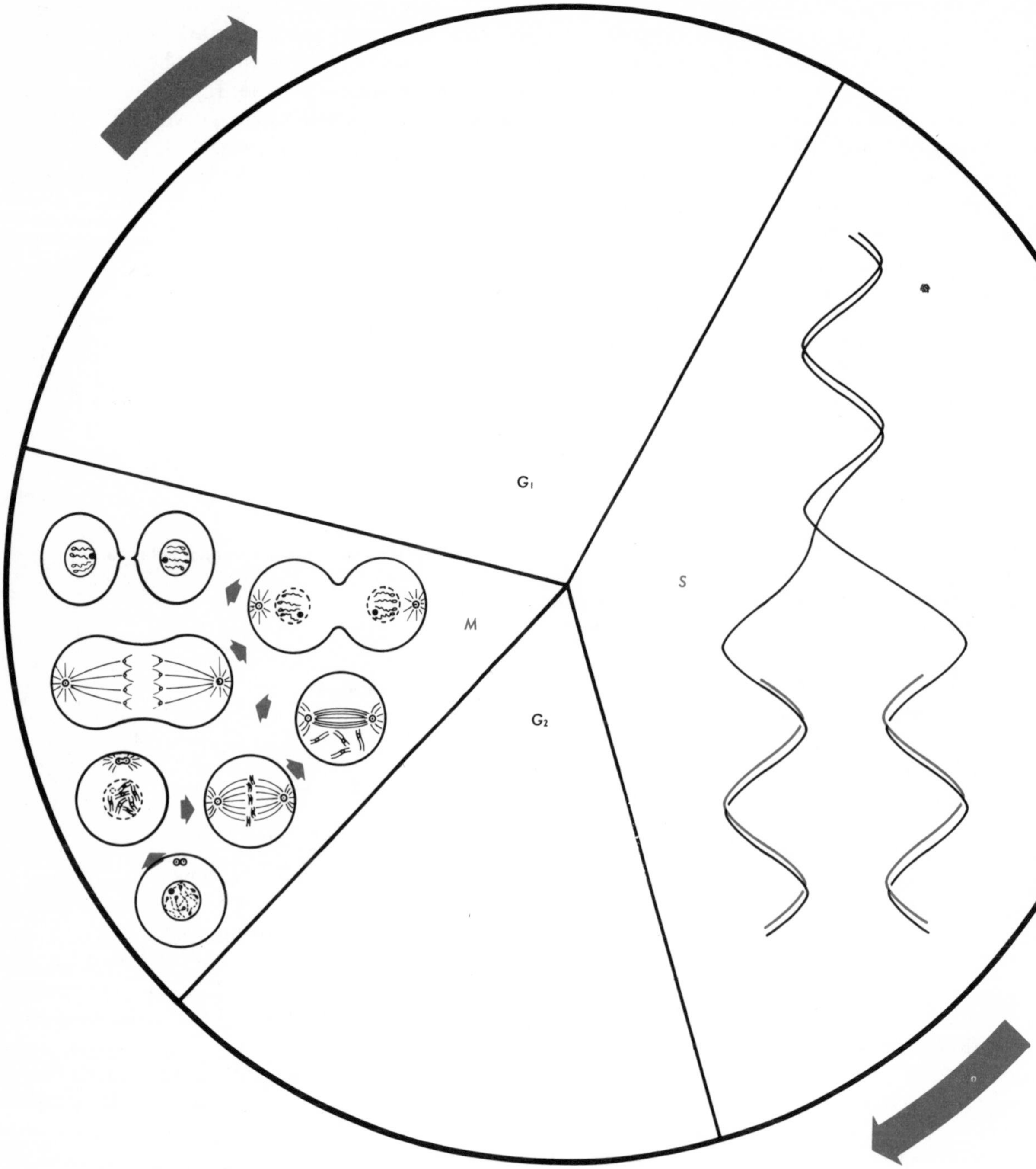

Fig. 5-7. The cell cycle.

from the endoderm and muscle is derived from the mesoderm. These three layers are composed of the four basic tissue types of the body: epithelium, connective tissue, muscle tissue, and nerve tissue. All four of these types can be present in the same organ. For example, the stomach wall is lined with epithelial cells, and contains connective tissue cells, muscle cells and nerve fibers.

Epithelium

Epithelial cells are found either in sheets, such as in the skin or stomach lining, or in glandular tissue. Squamous epithelial cells are flattened, cuboidal epithelial cells that are as tall as they are wide, and columnar epithelial cells are taller than they are wide (Fig. 5-8).

As seen in Fig. 5-8, epithelial cells are supported by the basement membrane, which separates them from underlying tissue. This basement membrane is not made of cells, but is believed to be secreted by the epithelial cells themselves.

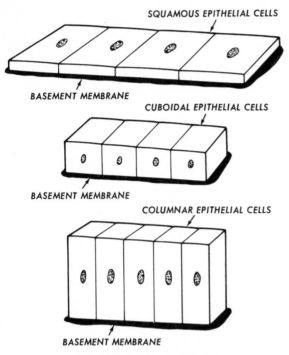

Fig. 5-8. **Types of epithelium.**

stance known as keratin, the material that also makes up the hair and fingernails. Keratin forms on external surfaces, such as the skin, and makes the epidermis more resistant to mechanical forces, such as rubbing. Other areas, the mouth and the vagina, for example, are lined with stratified

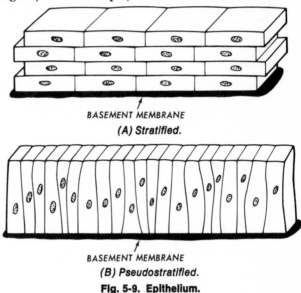

Fig. 5-9. **Epithelium.**

In most cases, the epithelium is made up of several layers, or strata, of cells. Fig. 5-9a shows an example of stratified epithelium. These cells are called stratified because only the bottom layer is in contact with the basement membrane. Fig. 5-9b shows pseudostratified epithelium. It is called pseudostratified because all the cells touch the basement membrane, but since the cells are of different sizes, the nuclei are at different levels, giving the impression of several layers.

Epithelial cells are suited to various roles in the body. Protection, a major role, is carried out mainly by stratified squamous epithelium, as in the epidermis. These cells, under certain circumstances, can form a horny layer made of a sub-

squamous epithelium, but there is normally no keratin layer present. Sometimes it is present in areas exposed to constant irritation. These squamous cells are held together tightly by desmosomes. The desmosomes in the superficial layers eventually weaken and the cells are sloughed off by friction. This phenomenon is known as exfoliation.

Another role played by epithelial cells is absorption. These cells are usually columnar epithelial cells, commonly found in parts of the digestive system.

The third major role played by epithelial cells is reproduction. Epithelial cells of the testis in the male undergo meiosis and produce sperm cells while other epithelial cells in the male produce sex hormones. In the female, the ovum is produced by epithelial cells. When the ovum is fertilized, specialized epithelial cells nourish and support the embryo.

Review Questions

1. Label the following "male" and "female" where appropriate

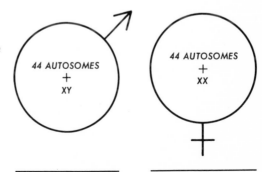

2. Match the following:
   a. Haploid
   b. Diploid

   _____ parents

   _____ offspring

   _____ sex cells (i.e., sperm or egg)

   _____ fertilized egg

3. Match the appropriate labels with the illustrations below:

   interphase   anaphase
   early prophase   early telophase
   prophase   late telophase
   metaphase

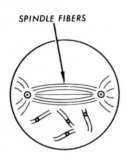

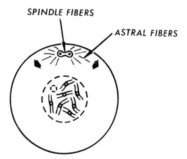

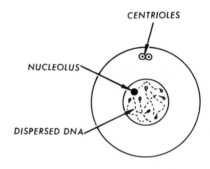

_____

_____

_____

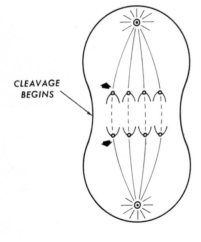

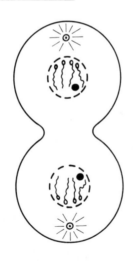

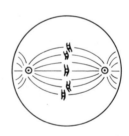

_____

_____

_____

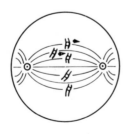

_____

4. In the illustrations below:

   a. Label: 23 ♂ = a

              23 ♀ = b

              33   = c

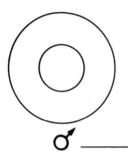  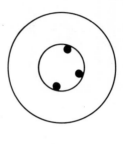

♂ _____   _____   _____

   b. Based on chromatin, indicate which cells are:

       a. female _____

       b. male _____

       c. Kleinfelter's syndrome _____

       d. XY chromosomes _____

       e. XX chromosomes _____

       f. XXXXY chromosomes _____

5. Match the appropriate labels with the illustrations below:

For illustration 1 below:

_____ stratified epithelium

_____ squamous epithelium

_____columnar epithelium

For illustration 2 below:

_____ cuboidal epithelium

_____ pseudostratified epithelium

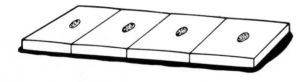

_____

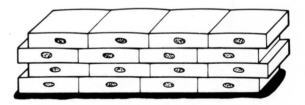

_____

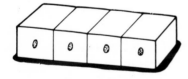

_____

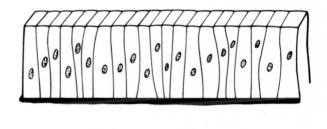

_____

_____

chapter 6

Female Genital Tract: The Ovaries

The male genital tract has several external parts, but the female gonads or ovaries are located inside the body. Fig. 6-1 schematically shows the overall relationship of the female parts. The vagina is the canal which leads from the exterior to the uterus. The uterus branches off into two "horns" which are attached to the fallopian tubes. Each fallopian tube curves around an ovary, and on each side, the broad ligament covers both ovary and tube. In general, the ovary produces the fe-

male germ cell, or ovum (egg), which passes into the fallopian tube where it may be fertilized. The fertilized egg usually travels down the tube and implants itself on the wall of the uterus where the fetus develops. If the egg is not fertilized, it merely passes to the outside of the body. We will discuss each part of the female genital tract, examining the anatomy and physiology and how the different parts are related to one another.

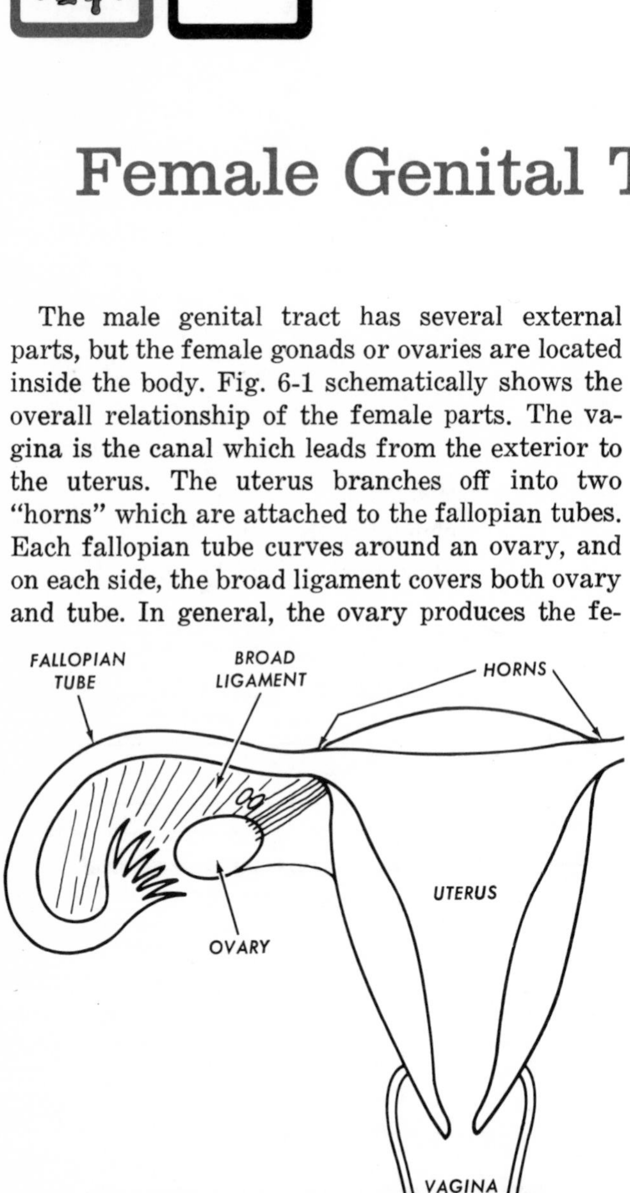

The Ovary

We shall begin with the ovary because it has a controlling influence on the rest of the genital tract. The ovary is an organ with two main functions: endocrine and exocrine. Exocrine means external secretion; that is, it secretes something outside the body. Endocrine literally translated means internal secretion. Endocrine glands, such as the thyroid and the pituitary, secrete hormones which induce effects on other regions of the body. Hormones travel through the bloodstream. The endocrine secretions of the ovary are the female sex hormones called estrogen and progesterone. The exocrine function of the ovary involves the release of the ovum (Fig. 6-2).

These functions are performed by a part of the ovary known as the follicle. Each follicle contains an ovum or egg cell, and is surrounded by cells which produce the sex hormones. The pituitary gland controls the development of the follicles which, in turn, control the rest of the genital tract.

Fig. 6-3 shows the general appearance of an ovary. There are two main regions, the cortex

Fig. 6-1. Schematic view of female genital tract.

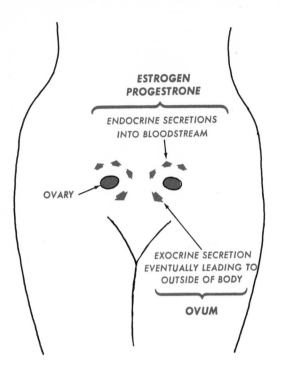

ESTROGEN
PROGESTRONE

ENDOCRINE SECRETIONS
INTO BLOODSTREAM

OVARY

EXOCRINE SECRETION
EVENTUALLY LEADING TO
OUTSIDE OF BODY

OVUM

Fig. 6-2. Dual functions of the ovaries.

and the medulla. The cortex is the outer layer and the medulla is the inner portion. The ovary is covered with a layer of simple, cuboidal epithelium, which is known as the germinal epithelium. Beneath the epithelium there is no basement membrane, but a layer of loose connective tissue. The medulla also is made up of connective tissue. Throughout the ovary are the follicles, seen in various stages of development. These follicles constitute the basic functional units of the mature ovary.

The developing fetus produces up to 400,000 ova, each one covered by a single layer of follicular cells. These primitive follicles, known as primordial follicles remain in a dormant state until they are stimulated by the follicle-stimulating-hormone, or FSH, of the pituitary gland. Each month one primordial follicle begins developing and releases an ovum. Obviously, all 400,000 follicles cannot possible be released in the average 30 year period from puberty to menopause, so some of them undergo a type of degeneration, known as involution (Fig. 6-4).

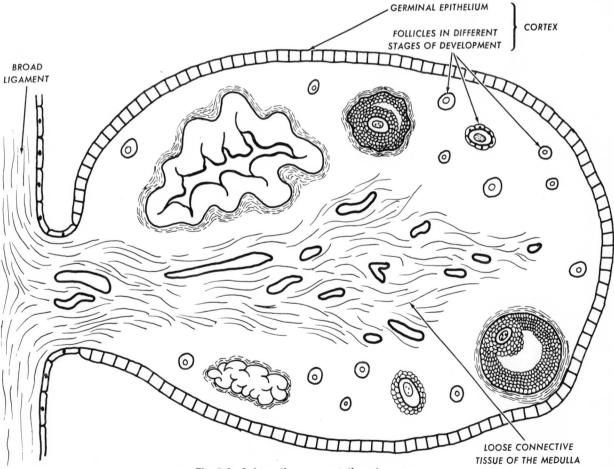

GERMINAL EPITHELIUM

FOLLICLES IN DIFFERENT
STAGES OF DEVELOPMENT

CORTEX

BROAD
LIGAMENT

LOOSE CONNECTIVE
TISSUE OF THE MEDULLA

Fig. 6-3. Schematic representation of ovary.

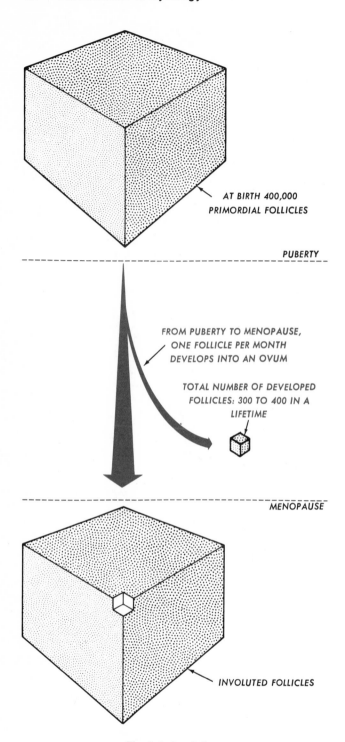

Fig. 6-4. Involution.

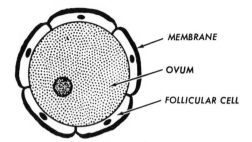

Fig. 6-5. Primordial follicle.

Follicular Development

Fig. 6-5 shows a primordial follicle, located in the ovarian connective tissue. It is composed of an ovocyte, surrounded by a layer of flat follicular cells, and enclosed by a membrane. The primordial follicle remains in this state until stimulated to begin the growth stage by becoming a primary follicle.

Fig. 6-6 shows a primary follicle. Notice that the ovocyte has grown, the follicular cells are cuboid and in several layers, and the surrounding membrane has thickened.

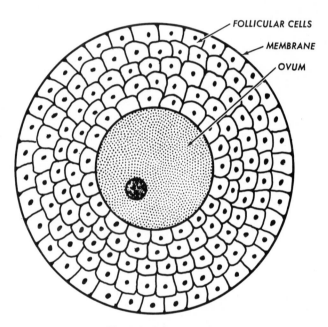

Fig. 6-6. Primary follicle.

Fig. 6-7 shows the next step, the growing follicle. A cavity, or antrum, appears at this stage and the follicular cells become even more numerous. Around the membrane there is a layer of cells, called the theca interna, and around that, a fibrous covering, the theca externa. These two additions will play an important role later. The follicular cells, called granulosa, secrete a liquid which fills the antrum. Note the prominent cumulus oophorus, a bulge which contains the ovum.

Fig. 6-8 depicts the next stage, when the structure is called the mature follicle or graafian follicle. The granulosa cells around the ovum take on a more ordered appearance and are known as the

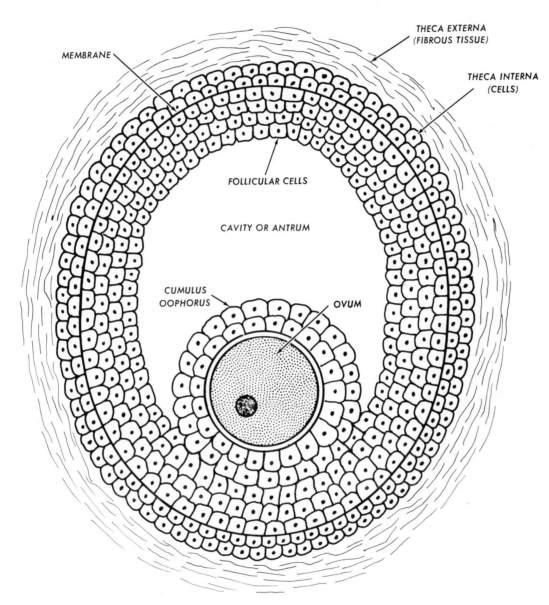

THECA EXTERNA
(FIBROUS TISSUE)

THECA INTERNA
(CELLS)

MEMBRANE

FOLLICULAR CELLS

CAVITY OR ANTRUM

CUMULUS
OOPHORUS

OVUM

Fig. 6-7. A growing follicle.

corona radiata. Soon, the follicular liquid begins to separate the cells of the cumulus oophorus (Fig. 6-9). At this stage, the follicle is at the peripheral limits of the ovary and appears as a lump on the surface. The entire follicular wall facing the outside of the ovary gets thinner and eventually the follicle ruptures, releasing the ovum through the hole, or stigma. This is ovulation (Fig. 6-10). The ovum is released and the remainder of the follicle undergoes a specialized degeneration.

After ovulation, the granulosa is invaded by vessels and the central region coagulates (Fig. 6-11). This mass is known as the corpus luteum. If the ovum is never fertilized by a sperm, the corpus luteum develops fully, with a maximum reached in 9 days, then begins to degenerate. The corpus luteum is an endocrine gland. The theca interna is sometimes called the thecal gland because it produces estrogens. Actually, estrogens are produced from the beginning by the growing follicle, but before ovulation, estrogen is the only hormone secreted by the ovary. After ovulation, estrogen continues to be secreted, but the hormone progesterone is also secreted by the luteal cells.

The final fate of the degenerating corpus luteum is its transformation into the corpus albicans, which is no more than a whitish scar on the ovary. Fig. 6-12 summarizes the process of ovulation and hormone secretion.

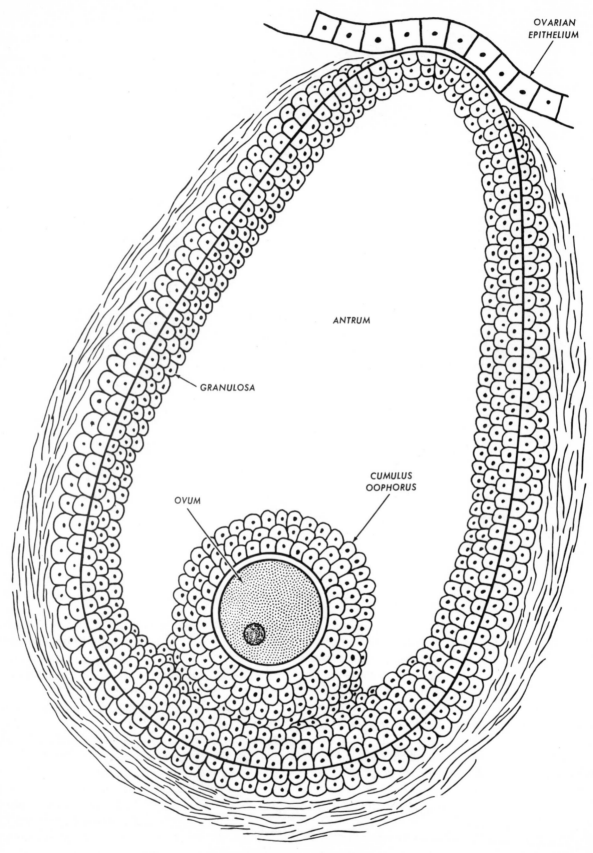

OVARIAN
EPITHELIUM

ANTRUM

GRANULOSA

CUMULUS
OOPHORUS

OVUM

Fig. 6-8. Graafian or mature follicle.

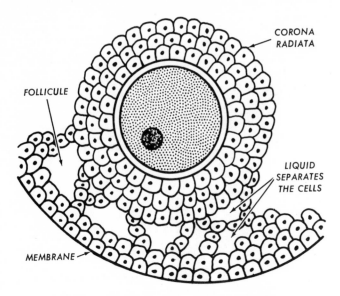

CORONA RADIATA

FOLLICULE

LIQUID SEPARATES THE CELLS

MEMBRANE

Fig. 6-9. Cumulus oophorus before ovulation.

A different process occurs if the egg is fertilized. The corpus luteum then is known as the corpus luteum of pregnancy and the endocrine processes are different.

Endocrine Control of Ovulation

The region of the brain known as the hypothalamus sends out basic commands to the pituitary gland (Fig. 6-13). The pituitary gland synthesizes and secretes FSH and LH (luteinizing hormone), both classified as gonadotropic hormones. The pituitary gland, however, must wait for the hypothalamus to send out releasing factors which permit the pituitary to release its hormones. These releasing factors are specific to the hormones and allow the hormones to be released when needed, as shown in Fig. 6-14.

At the beginning of the cycle, the FSH-releasing factor secreted by the hypothalamus allows the pituitary to secrete FSH, which initiates the follicular changes (remember, only one follicle can mature at a time). Next, the LH-releasing factor causes the LH to be secreted. LH conditions the ovary for ovulation and later induces the corpus luteum to develop. In summary: Before ovulation, only estrogen is secreted, while after ovulation both estrogen and progesterone are secreted (Figs. 6-15, 6-16).

Drug Controlled Ovulation

Ovulation can be induced by simulating the pituitary secretions. This can be accomplished by injecting drugs with properties similar to FSH and LH. Human menopausal gonadotrophin (HMG) has an FSH-like action, while human chorionic gonadotrophin (HCG) has an LH-like action (Fig. 6-17). Both FSH and LH are classified as gonadotrophins because they have a particular effect on the gonads.

In recent years, contraceptives which inhibit ovulation have been devised. In general, these drugs are artificial estrogens and progesterones which block the gonadotrophic hormones. Actually, they trick the pituitary into thinking that no more gonadotrophins are needed, and, since the pituitary is self-regulating, no gonadotrophins are released and no follicle can develop to produce an egg. (Fig. 6-18).

Estrogens

Estrogens are responsible for the development of the female genital tract. The role played by these hormones is demonstrated in castrated subjects. If both ovaries are removed from a subject who has not yet reached puberty, the remaining parts of the genital tract will never reach maturity. If a woman of childbearing age loses both ovaries, the following changes will occur:

1. The vaginal epithelium will be reduced to several layers and no more cyclic variations will take place.
2. The uterus will reduce in size or atrophy.
3. Muscles and epithelium of the fallopian tubes will regress.
4. Auxiliary sexual glands such as the mammary glands and Bartholin's gland will atrophy (the clitoris is not affected).

If estrogens are supplied to these patients in the proper doses, the above effects will reverse themselves. Conversely, if estrogens are administered to a normal female of childbearing age, the fallopian tubes and uterus will become congested and the growing follicles and corpus luteum will degenerate.

Estrogens have a particular effect on general body metabolism: they favor the deposition of subcutaneous fat and lower blood cholesterol levels. This is one reason why women are less susceptible to atherosclerosis than men are.

Progesterone

While estrogens develop the genital tract, progesterone prepares the tract for pregnancy. Pro-

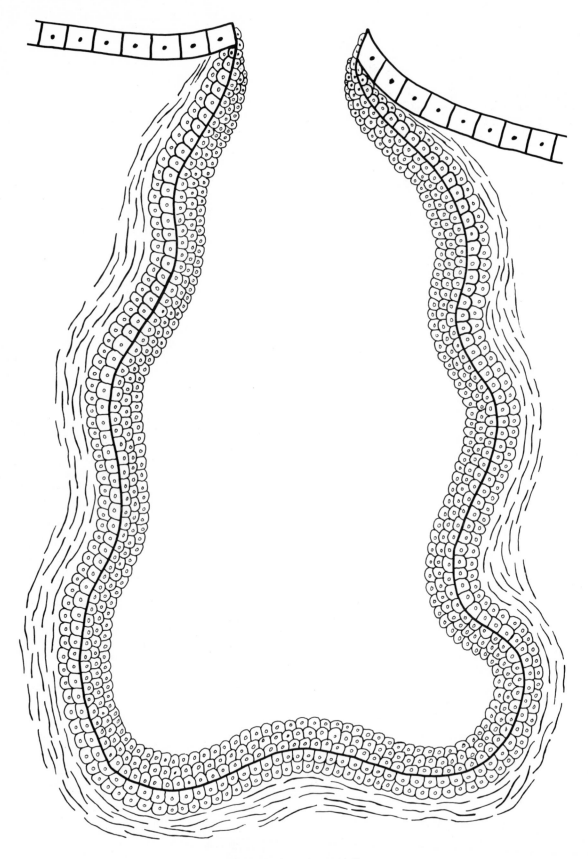

Fig. 6-10. Just after ovulation.

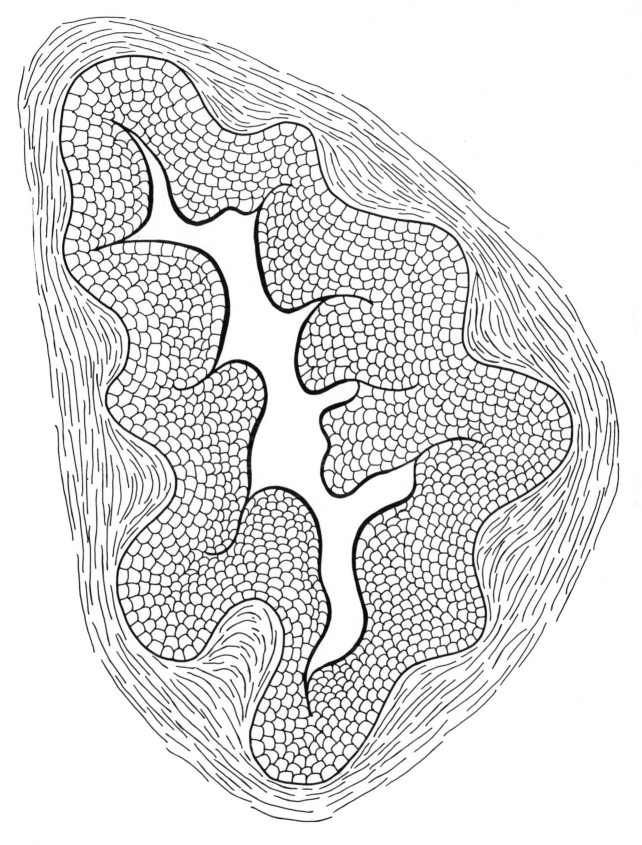

Fig. 6-11. Corpus luteum.

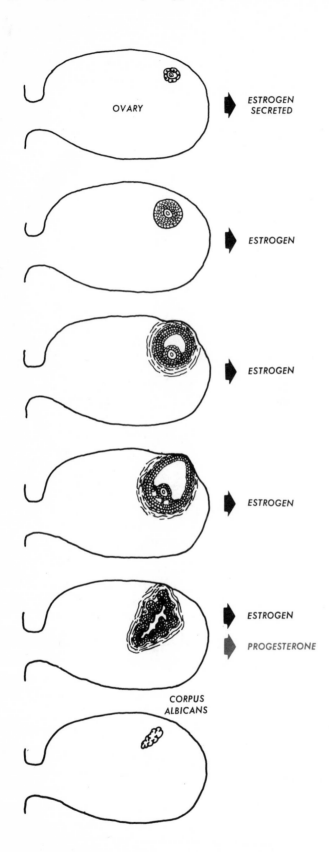

Fig. 6-12. Simplified summary of ovulation and hormone secretion.

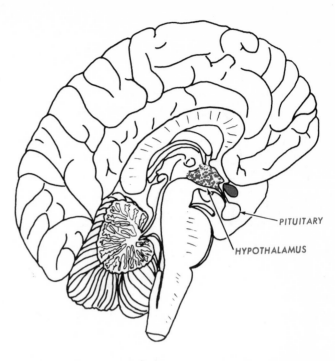

Fig. 6-13. Side view of brain to show location of pituitary gland.

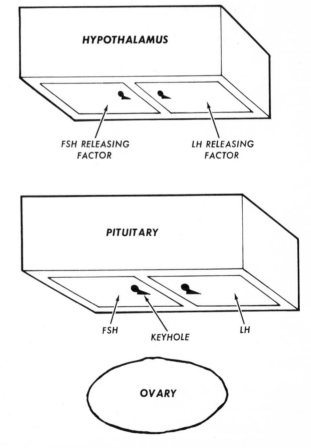

Fig. 6-14. Hypothalamus and pituitary activity before puberty. Before puberty, the ovary is full of primordial follicles, but no FSH is released, therefore, no follicle development can take place and no sex hormones are released.

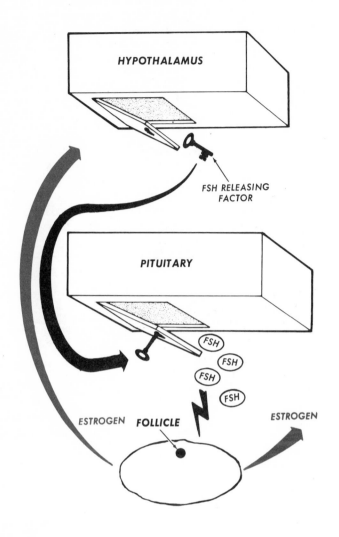

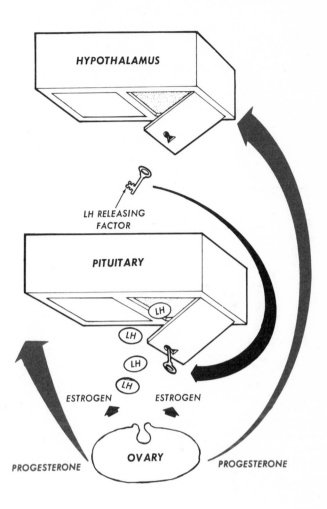

Fig. 6-15. Hypothalamus and pituitary activity during childbearing age (estrogen phase). After puberty and until menopause, cycle begins with FSH releasing factor which releases FSH into the bloodstream. The FSH stimulates follicle development and the follicle secretes estrogen. When blood estrogen reaches a certain level, it signals the hypothalamus, which in turn limits the amount of FSH released.

Fig. 6-16. Hypothalamus and pituitary activity during childbearing age (progesterone phase). At this point, estrogen levels drop and LH is released by hypothalamus. The LH causes the ruptured follicle to secrete progesterone, which is controlled by the hypothalamus by feedback.

gesterone secretion follows estrogen secretion and works with estrogen in the second half of the cycle (Fig. 6-19). This hormone prepares the uterus for ovo-implantation and inhibits the spontaneous contractions of the myometrium to prevent dislodging the fetus. The fact that progesterone increases body temperature has been used as a sign of ovulation.

Estrogen and progesterone work together: estrogen prepares the receptors for progesterone.

If progesterone alone is secreted, the effects are quite different and inefficient, so there is a cooperation or synergism between the two sex hormones.

Menarche

This is the beginning of the first menstruation. Gonadotrophic hormones are first secreted and the ovaries produce a mature egg for the first

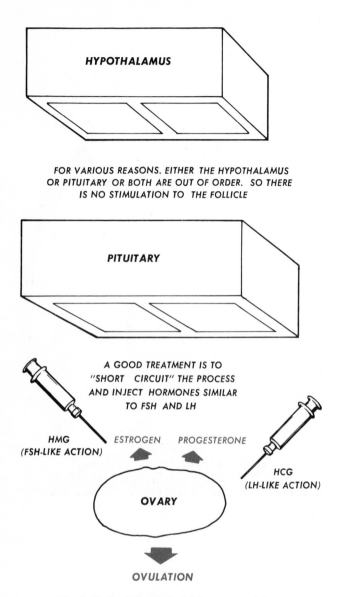

Fig. 6-17. Ovulation induced by means of drugs.

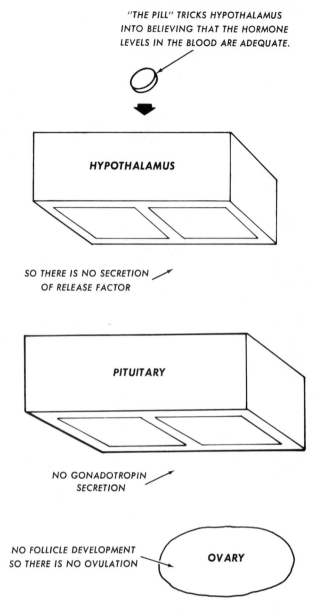

Fig. 6-18. Ovulation suppressed by means of drugs.

time. Menarche marks the beginning of the child-bearing period of life.

Menopause

The menopause is the end of the childbearing age. The ovary appears less receptive to the gondotrophins and the number of growing follicles decreases until all the follicles degenerate.

Because of the drop in hormone secretion, the genital tract and the breasts atrophy. The pituitary gonadotrophin secretion is turned off when sex hormone levels rise, and, since the ovary has stopped estrogen and progesterone production, the pituitary secretes large amounts of FSH with no results (Fig. 6-20). No follicles develop, and eventually no more progesterone is secreted.

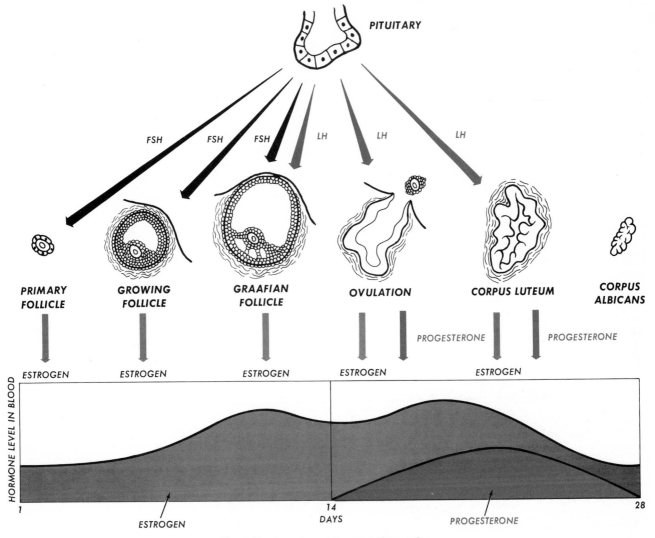

Fig. 6-19. Overview of the ovulation cycle.

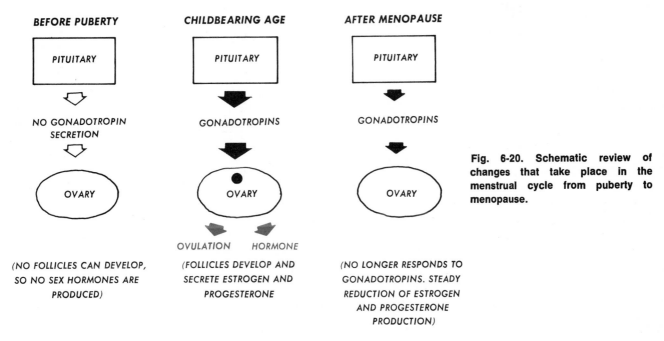

Fig. 6-20. Schematic review of changes that take place in the menstrual cycle from puberty to menopause.

Review Questions

1. In the illustration below, label:
   a. Uterus
   b. Fallopian tube
   c. Broad ligament
   d. Vagina
   e. Ovary
   f. Horns

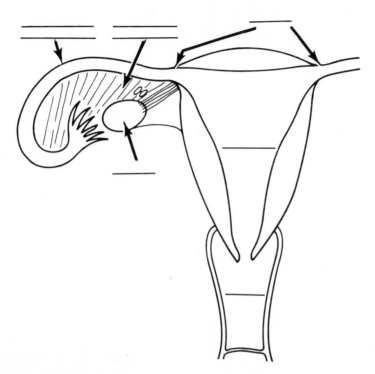

2. The endocrine function of the ovary releases the hormones _____ and _____.
3. The exocrine function of the ovary releases an _____.
4. In the illustration below, label:
   a. Germinal epithelium
   b. Medulla
   c. Cortex
   d. Follicles
   e. Broad ligament

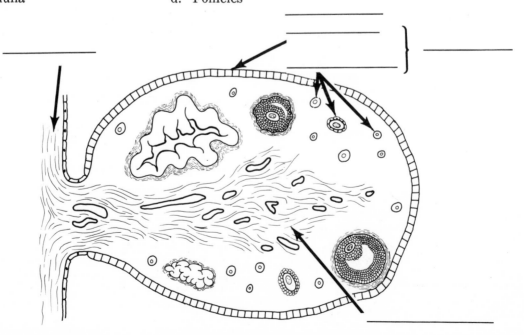

5. For the illustration below:

   Title _____

   Label: membrane, ovum, follicular cells.

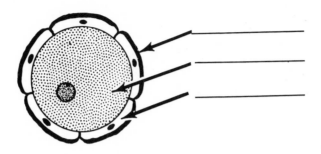

6. For the illustration below:

   Title _____

   Label: membrane, ovum, follicular cells.

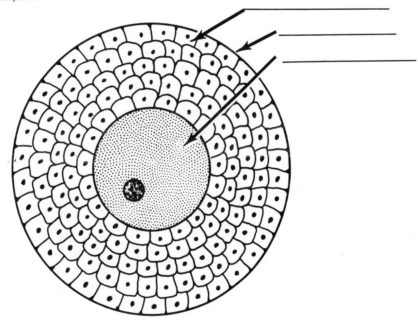

7. For the illustration below:
   Title _____

   Label: antrum, follicular cells, ovum, theca interna, theca externa

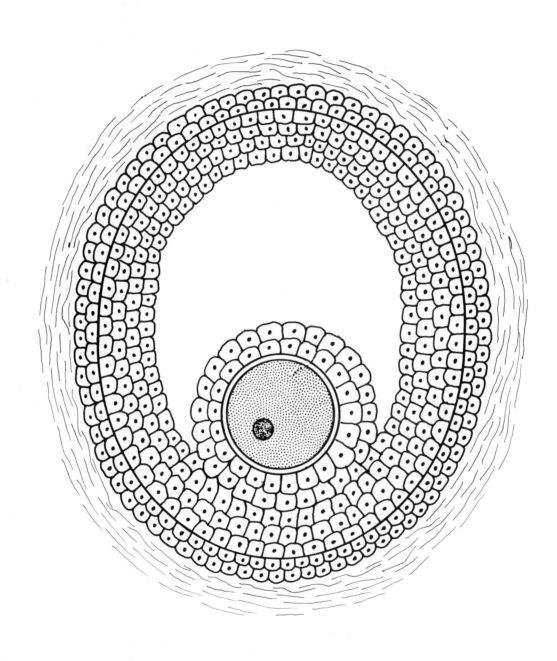

8. For the illustration below:

Title _____

Label: germinal epithelium, antrum, granulosa, ovum, theca interna, theca extern, cumulus
oophorus, corona radiata, follicular liquid, membrane

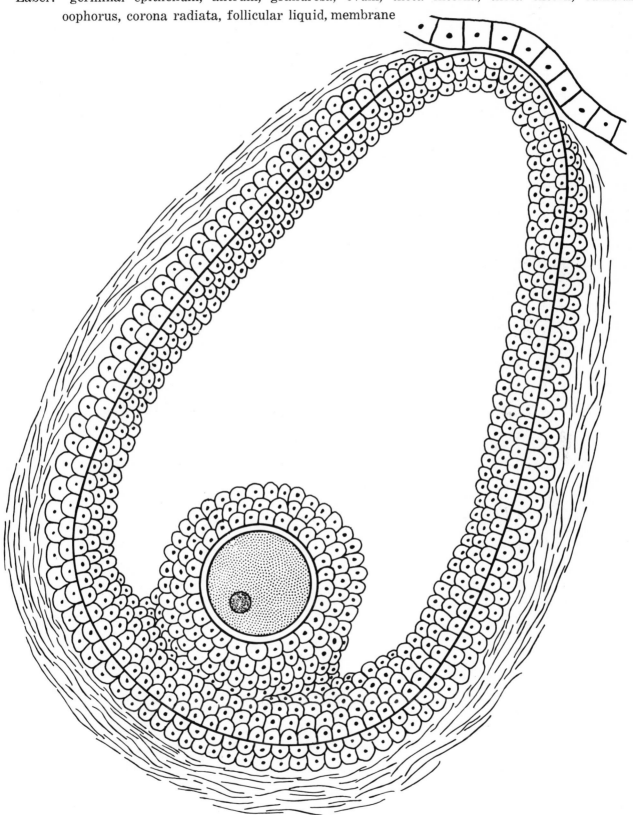

9. a. What does the diagram below represent?   b. Where is the stigma?

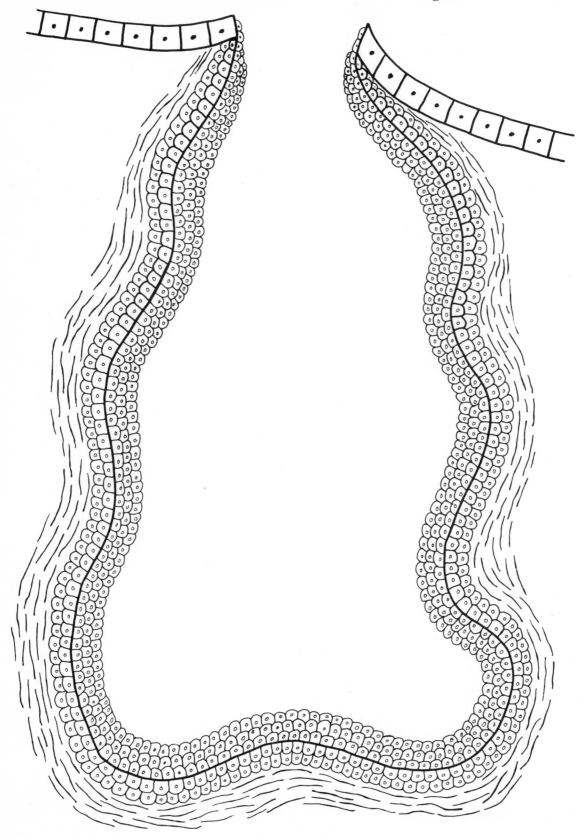

10. Identify the structure shown in the illustration below.

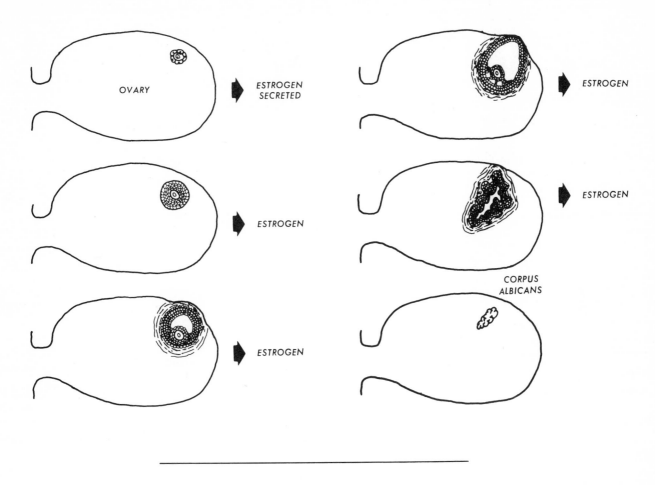

11. At which stages shown in question 10 is estrogen produced?

12. At which stages shown in question 10 is progesterone produced?

13. Match the appropriate term from the left column with that in the right column:
    a. FSH
    b. LH
    c. releasing factors
    d. estrogen
    e. progesterone
    f. pituitary gland
    g. prepares genital tract
       for pregnancy each cycle
    h. responsible for develop-
       ment of female tract
    i. acts like FSH
    j. acts like LH

    _____ Human menopausal gonadotropin
    _____ Human chorionic gonadotropin
    _____ Estrogen
    _____ Progesterone
    _____ Pituitary hormones
    _____ secreted by the hypothalamus
    _____ directly stimulates estrogen secretion
    _____ under direct influence of hypothalamus
    _____ directly stimulates progestone secretion
    _____ gonadotropins

14. True or False
    _____ FSH is secreted before LH in each cycle.
    _____ Menopause is the end of the childbearing phase of life.
    _____ In menopause, FSH secretion suddenly stops.

The Female Genital Tract: Fallopian Tubes and the Uterus

A fallopian tube is represented schematically in Fig. 7-1, which shows the infundibulum, ampulla, and isthmus. This entire structure, as well as the

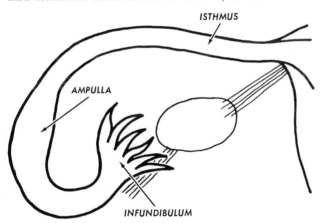

Fig. 7-1. Fallopian tube.

ovary, is covered with the broad ligament. Fig. 7-2 is a cross sectional view of the fallopian tube. Note the mucosa, the epithelial lining, and its enormous surface area. The muscular layer is made up of circular and longitudinal smooth (involuntary) muscle fibers. The entire organ is covered by the outer layer called the serosa.

The fallopian tubes have three biological functions. They are:

1. Capture of the egg (ovum) after ovulation
2. Transport of the egg to the uterus
3. "Capacitation" of sperm to fertilize egg.

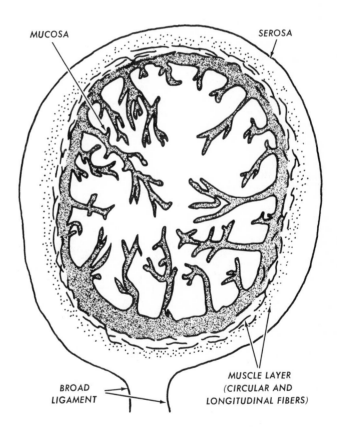

Fig. 7-2. Cross section of fallopian tube.

1. Capture of the Egg. The infundibulum of the fallopian tube, which covers the ovary during ovulation, is relatively mobile compared with the rest of the tube. If the egg released by the ovary does not pass directly into the mouth of the tube,

it is drawn in by the fluid flowing into the tube (Fig. 7-3). Fluid constantly is being drawn in through the fallopian tube and absorbed in the epithelial folds of the interior. This constant absorption causes an actual flow of fluid into the infundibulum portion. Since almost any of the thousands of follicles can develop into an ovulating follicle, and since they break through the surface of the ovary at the nearest point, there is no guarantee that the egg will fall directly into the infundibulum. However, with a continuous fluid current leading into the tube, the eggs which exit through the part of the ovary facing away from the opening of the tube are drawn in.

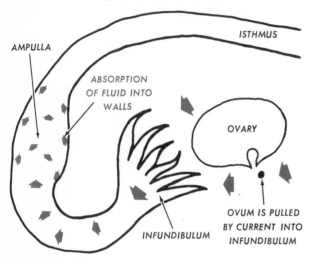

Fig. 7-3. Attraction of egg into fallopian tube by fluid currents.

2. Transport of the Egg to the Uterus. After the egg enters the tube, the current can move it only so far, since the liquid eventually is absorbed and does not travel all the way to the uterus. Actually, most of this movement is provided by the ciliary movements of the epithelial cells that line the passage (Fig. 7-4). Cilia are cytoplasmic hair-like extensions which wave in a given direction. In the fallopian tubes, these ciliated cells move the egg toward the uterus. It takes about three days for a fertilized egg to reach the uterus.

3. Fertilization of the Egg. The egg is fertilized in the fallopian tube and then migrates to the uterus where it is implanted onto the uterine wall. In order for a sperm to fertilize an egg, it must penetrate the egg (the egg is many times larger than a sperm). Therefore, sperm must pass through the uterine tube and be conditioned by certain factors which "capacitate" it, that is, ren-

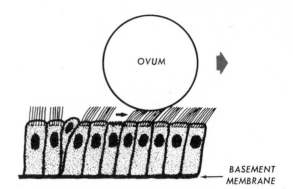

Fig. 7-4. Movement of ovum across fallopian epithelium.

der it capable of penetrating the egg. It is not yet known exactly what the tube does to the egg, but the tube plays an important part in preparing the sperm for penetration. Once the egg is fertilized, it begins to divide, even while it is still traveling down the tube.

The Fallopian Tube During the Endocrine Cycle

Before ovulation, the epithelium is thin and the ciliated cells are present in very small proportion. At ovulation the epithelium is thick and numerous ciliated cells are present. After ovulation, the epithelium thins, but the ciliated cells remain active for awhile, then begin to decrease in number.

The development of the fallopian tubes depends on the endocrine factors estrogen and progesterone, demonstrating the close relationship between the parts of the genital tract and the hormonal messengers.

The Uterus

The uterus is an organ which plays an essential role in reproduction. The fertilized egg is implanted, or attached to the uterine wall and nourished and supported through the months until birth. The fertilized egg may develop in the fallopian tube, but this is abnormal.

Fig. 7-5. shows the relative location of the uterus in frontal and profile views. The uterus is located above the vaginal passage and receives the two fallopian tubes laterally.

Fig. 7-6 is a schematic view of the uterus, identifying the main regions. The corpus uteri, or body of the uterus, is the main region. It narrows into the isthmus uteri which then leads into the cervix, or neck, of the uterus. All these regions are made up of three layers: the endometrium,

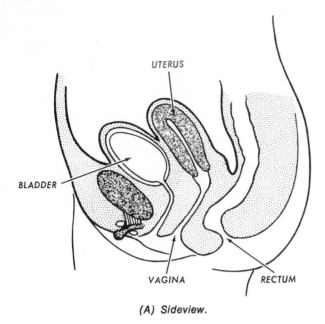

(A) Sideview.

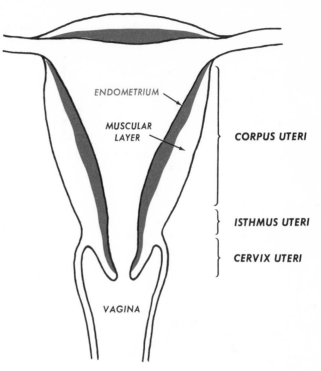

Fig. 7-6. Parts of the uterus.

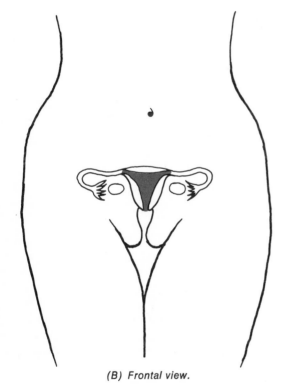

(B) Frontal view.

Fig. 7-5. Location of the uterus.

or mucosa; the myometrium, or muscular layer; and the serosa.

The endometrium is the thickest in the corpus uteri region, and is composed of two zones, the superficial layer and the basal layer. Fig. 7-7 shows that the endometrium is lined with a simple columnar epithelium and a stroma. This stroma contains numerous glands and vessels such as

arterioles (small diameter arteries), venules (small diameter veins), and lymph passages. (As you can see in Fig. 7-7 most of the vessels are located in the residual layer.) The endometrium changes in a cyclic manner as the endocrine cycle progresses. The thickness of the endometrium varies with the phase of life. Before puberty and after menopause, the endometrium is much thinner than during the childbearing years. The endometrium of the isthmus is similar to the endometrium of the corpus uteri, but it is thinner and has a less important function in pregnancy.

The cervix is richly endowed with mucus secreting glands which empty directly into the cervical canal (Fig. 7-8). These glands are active only during the childbearing years. The cervix will be discussed further in the section dealing with the vagina.

The Myometrium

The myometrium, or muscular layer, is made up of three layers of muscle. As Fig. 7-6 shows, the middle layer is the thickest. The muscular layer is richly supplied with blood vessels, which nourish the developing fetus. These blood vessels each are surrounded by special muscle fibers which can contract and block the blood flow like microscopic tourniquets.

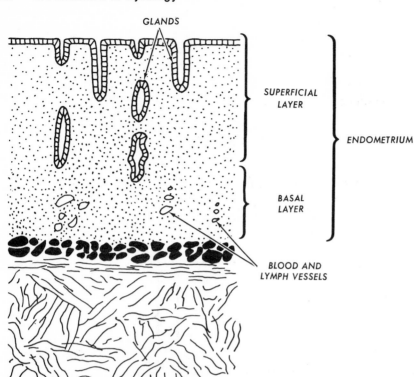

GLANDS

SUPERFICIAL
LAYER

ENDOMETRIUM

BASAL
LAYER

Fig. 7-7. Cross section of the uterus wall.

BLOOD AND
LYMPH VESSELS

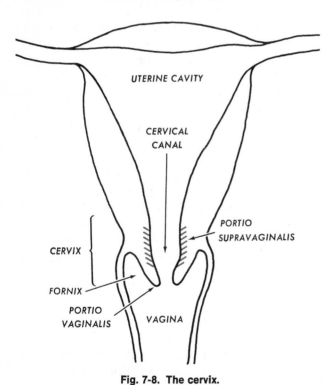

UTERINE CAVITY

CERVICAL
CANAL

PORTIO
SUPRAVAGINALIS

CERVIX

FORNIX

PORTIO
VAGINALIS

VAGINA

Fig. 7-8. The cervix.

Physiology of the Uterus

This section discusses cervical gland secretions, uterine contractions, and the menstrual cycle and how they relate to the ovulation cycle.

Cervical gland secretions consist mainly of mucus, a slimy, thick liquid consisting of proteins. These secretions serve two purposes: to protect the uterus from infection, which may originate in the vagina; and to increase the sperm's chances of surviving the passage from the vagina into the uterus. Since the cervix is the "doorway" to the uterus, it must block the passage of bacteria into the uterus. The cervical mucus contains antibacterial substances, known as lysozymes, similar to the digestive enzymes found in lysosomes. The mucus also makes sperm passage easier during the fertile part of the female cycle. Since sperm cells require an alkaline environment for survival, the cervical mucus is alkaline during the fertile period and acid at the beginning and at the end of the cycle (Fig. 7-9). Therefore, like the rest of the genital tract, the cervix is controlled by the hormones secreted by the ovarian follicle.

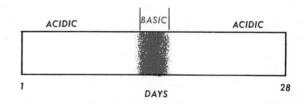

ACIDIC   BASIC   ACIDIC

1   DAYS   28

Fig. 7-9. pH of cervix during cycle.

Uterine Contractions

In the gravid, or pregnant, uterus, there are few spontaneous contractions. When labor begins, however, the myometrium contracts in regular waves which increase in frequency as the moment of birth approaches.

Menstruation

The bloody fluid, known as the menses, which is discharged from the uterus during menstruation is composed of uncoagulated blood, mucus, endometrial secretions, and cellular debris. The average menstrual cycle occurs approximately every 25 to 32 days, although other variations are possible (Fig. 7-10). The cycle is dependent on the ovulatory cycle and is controlled directly by the ovarian hormones. The actual duration of menstruation usually varies from 3 to 6 days, with the maximum flow on the second or third day. Volume of flow can vary from 10 to 130 milliliters, but the mean volume is about 50 milliliters. After the age of 35, the volume usually decreases.

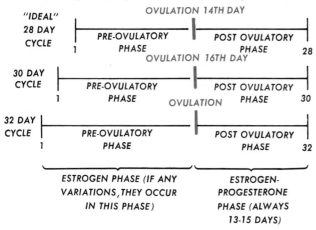

Fig. 7-10. Variations in the menstrual cycle.

Menstruation results from the breakdown of the superficial endometrium, which is caused by the lack of hormones. This process takes several days because the breakdown occurs in patches. The superficial layer of the endometrium prepares itself for the arrival of a fertilized egg at the beginning of each cycle. If the egg is not fertilized and implanted on the uterine wall, the superficial endometrium breaks down, leaving only the basal layer (Fig. 7-11). At the beginning of the next cycle, the basal layer will give rise to a new superficial layer, continuing the cycle.

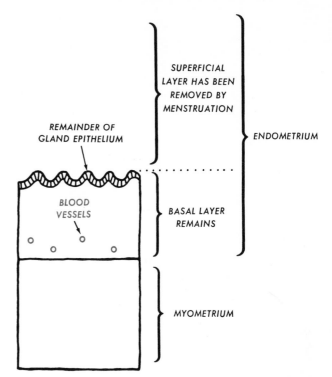

Fig. 7-11. Bare basal layer remains after previous menstrual period.

There are two phases in the menstrual cycle: the pre-ovulatory phase and the post-ovulatory phase. Theoretically, ovulation occurs on the 14th day of a cycle, but it usually is not this regular. Variations commonly occur, but no matter how long the cycle, the post-ovulatory phase takes 13 to 15 days (Fig. 7-10). Since we are discussing the effects of the two sex hormones on the female genital tract, we will refer to the pre-ovulatory phase as the estrogen phase and the post-ovulatory phase as the estro-progesterone phase.

The menstrual cycle can be divided into four stages: the menstrual stage, where there is external menstrual discharge; the proliferative (or follicular) stage, during which follicular growth and estrogen secretion occur; the progestational (or luteal) stage, associated with an active corpus luteum; and the ischemic (or premenstrual) stage, where blood flow through the coiled arteries is interrupted.

Fig. 7-12 shows the beginning of the proliferative (or follicular) stage of the endometrium. During this phase, the endometrium, under the influence of estrogen, undergoes a rapid regeneration. The epithelium, lost during menstruation, is regenerated and consists of a simple layer of cuboid cells, which later become columnar. This epithelium lines the glands, simple tubes that bur-

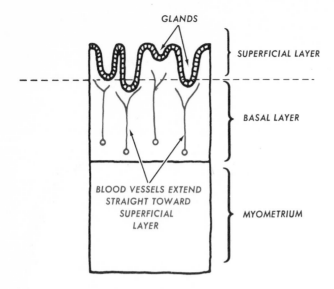

Fig. 7-12. Proliferative phase (early).

row into the endometrium. (During menstruation, these glands were lost except for the blind ends which extended into the untouched basal layer.) Vessels develop from the remains left in the basal layer and extend upward into the superficial layer. Fig. 7-13 depicts later developments in the proliferative stage. The mucosa has become much thicker, and will measure about 3 millimeters by the time ovulation occurs. The glands, straight until now, begin to take on a slightly wavy appearance, while the arterioles have taken on a corkscrew, or helical, appearance, but do not yet extend to the surface. Only capillaries and venules are present in the most superficial part of the endometrium.

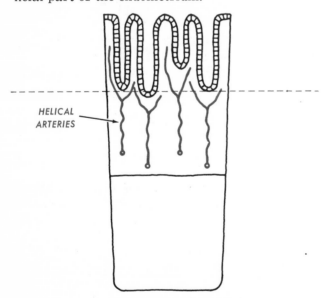

Fig. 7-13. Proliferative stage (late).

At this stage, ovulation occurs. If the ovum is not fertilized by a sperm, the corpus luteum forms. At this time, both estrogen and progesterone are being secreted.

Fig. 7-14 shows the progestational or luteal stage of the endometrial cycle. About forty-eight hours after ovulation and corpus luteum formation, the glands, which now are quite numerous, undergo changes. They become long and shaped like corkscrews. The inset of Fig. 7-14 shows how these glands have formed glycogen, a form of

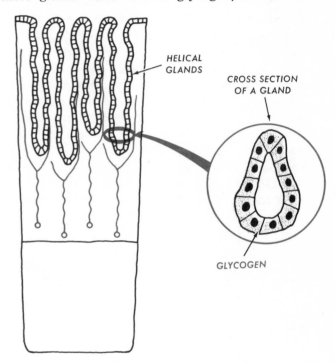

Fig. 7-14. Progestational or luteal stage.

storage for glucose. The vessels continue to grow and the endometrium begins to swell with edema, a condition where fluids leave the vessels and infiltrate the tissues. This swelling is considered to be in preparation for the implantation of the fertilized ovum. Later in the progestative stage (Fig. 7-15), the glands secrete abundant quantities of glycogen.

Fig. 7-16 shows the ischemic or premenstrual stage of the endometrium. This occurs 13 to 14 days after ovulation, and is characterized by vascular changes. The muscle fibers around the coiled arteries constrict and blood cannot reach the superficial layers. As a result, the tissues suffer from insufficient oxygen (anoxia) and begin to die.

Fig. 7-17 shows the menstrual stage of the endometrium. The superficial layers undergo ne-

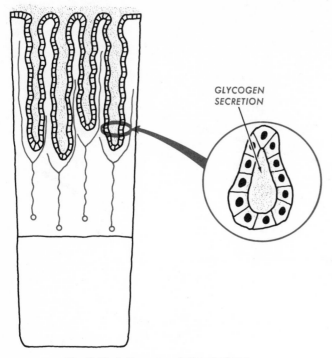

GLYCOGEN SECRETION

Fig. 7-15. Later progestational phase.

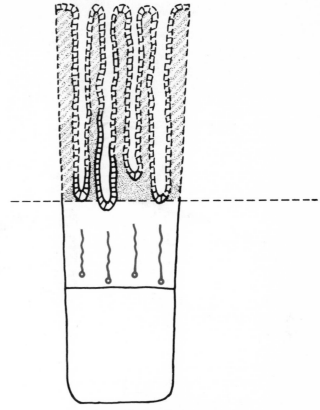

Fig. 7-17. Menstrual stage.

crosis or cell death. After an artery has been constricted for a certain time, it opens again and blood flows to the necrotic zone, separating the patches of dead tissue, which are shed with a little blood. Eventually, all the superficial layer is gone and only the raw basal layer and the blind ends of the glands remain. At this point, a new cycle begins. See Fig. 7-18 for a summary of the entire cycle and its relation to ovarian secretion.

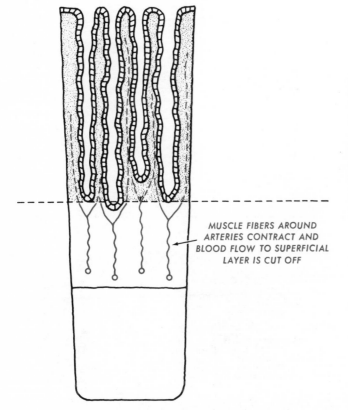

MUSCLE FIBERS AROUND ARTERIES CONTRACT AND BLOOD FLOW TO SUPERFICIAL LAYER IS CUT OFF

Fig. 7-16. Premenstrual stage.

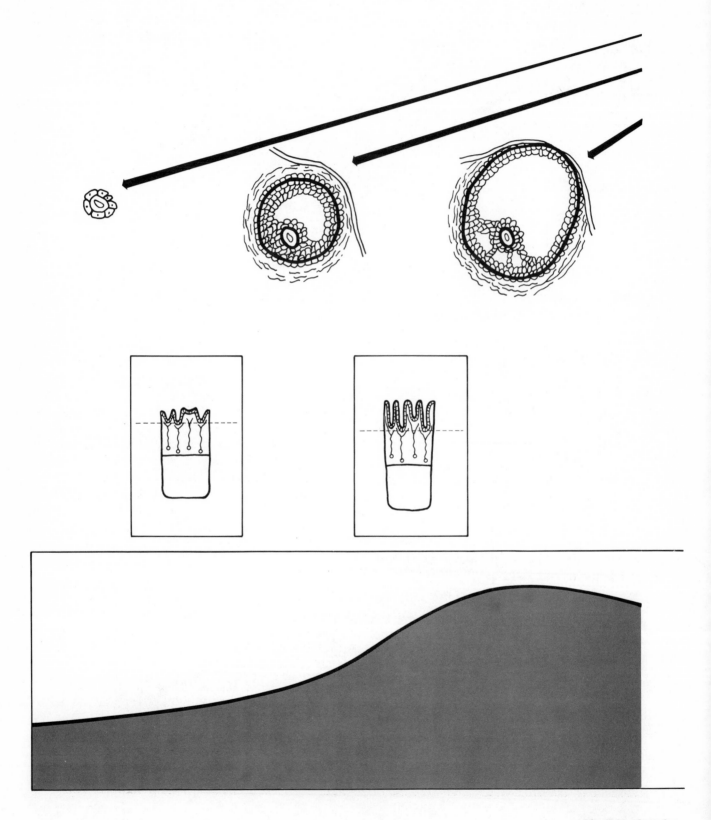

Fig. 7-18. Overview

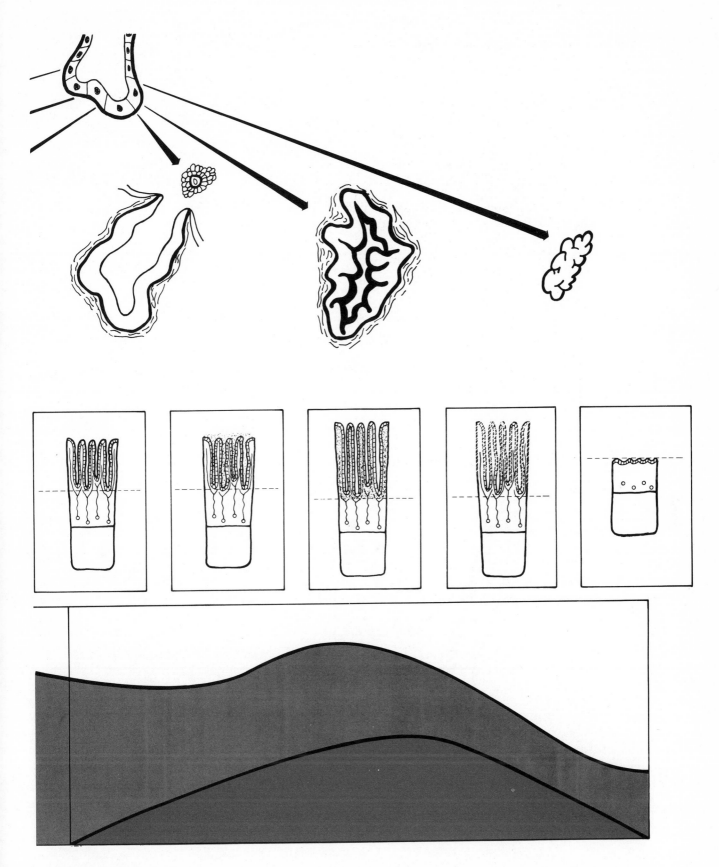

of menstrual cycle.

Review Questions

1. In the illustration below, label: Uterus, bladder, rectum, vagina

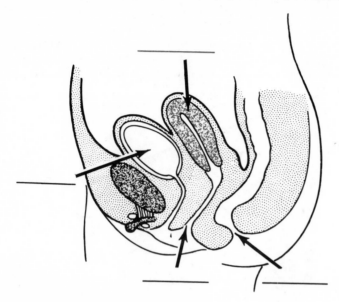

2. In the illustration below, label: Endometrium, muscle layer or myometrium, isthmus, cervix, endo-cervical canal, fornix, portio vaginalis.

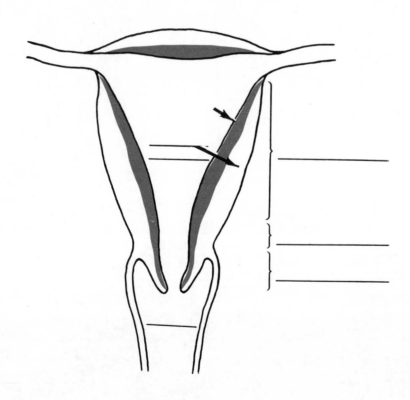

3. In a 28 day menstrual cycle, ovulation occurs on approximately the _____ day.
   In a 32 day menstrual cycle, ovulation occurs on the _____ day.

4. Match the illustrations below to the appropriate stage

   1. _____
   2. _____
   3. _____
   4. _____
   5. _____
   6. _____

   _____ a. menstrual stage
   _____ b. progestational or luteal stage
   _____ c. pre-menstrual stage
   _____ d. post-menstrual stage
   _____ e. proliferative stage
   _____ f. late progestational stage

   _____
   (See additional figures on Page 70.)

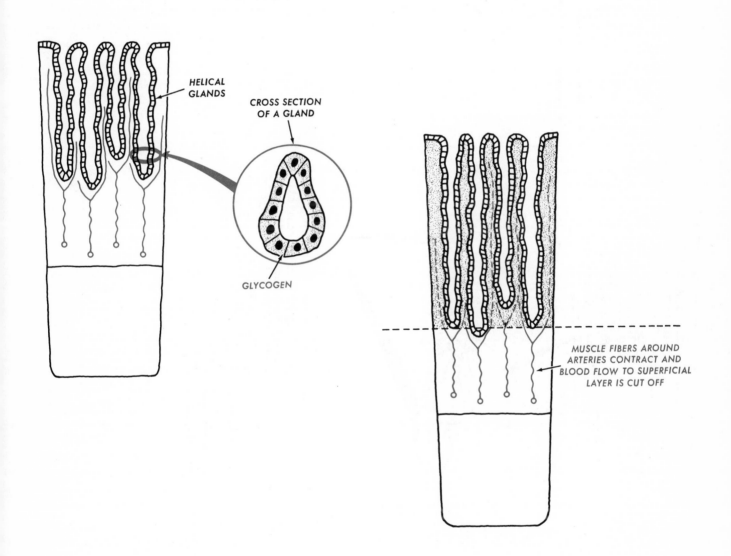

HELICAL GLANDS

CROSS SECTION OF A GLAND

GLYCOGEN

MUSCLE FIBERS AROUND ARTERIES CONTRACT AND BLOOD FLOW TO SUPERFICIAL LAYER IS CUT OFF

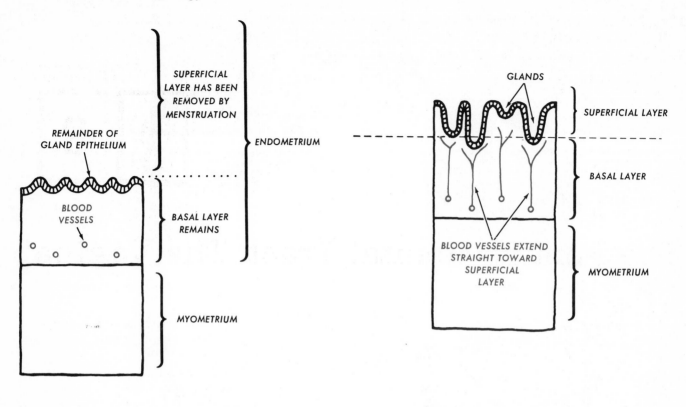

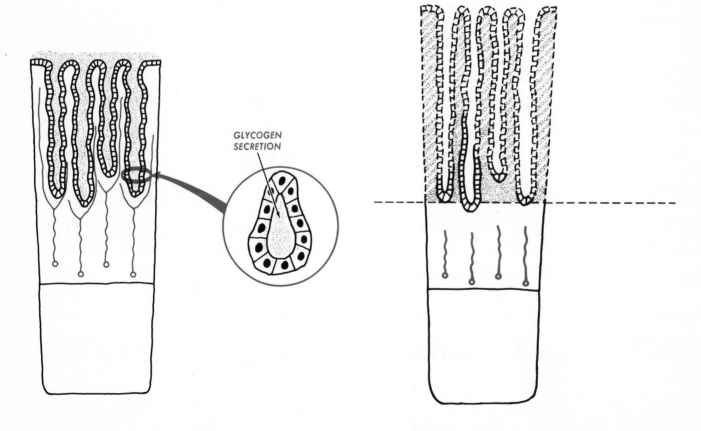

The Female Genital Tract: The Vagina

As shown on Figs. 8-1 and 8-2, the vagina is a tunnel-like structure which leads from the vulva to the uterus. It is in close proximity to the rectum, which is behind it, and to the bladder, which lies in front. The proximity of these two organs makes it possible for cancers to spread from the rectum and the bladder to the vagina and vice versa.

The cervix extends down into the vaginal cavity and is called the portio vaginalis at this point. Fig. 8-1 shows the area bounded by the cervix and the end of the vaginal pouch. This region, called the fornix, is the site of accumulated secretions from the uterine glands and exfoliated epithelial cells.

Fig. 8-1. Profile of internal genitals.

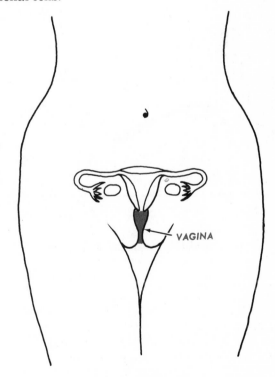

Fig. 8-2. Location of vagina.

The vaginal wall is smooth except for transverse ridges and two longitudinal swellings. These protrusions are made up of epithelial thickening. The wall consists of three layers: the mucosa, the muscular layer, and the serosa.

Vaginal Flora

The natural flora found in the vagina consists mainly of bacteria known as *Bacillus vaginalis*. This bacterium is nonpathogenic (except in certain cases) and belongs to the *Lactobacillus* group. The organism is polymorphic, (it can take several shapes), but usually appears as Gram-positive rods about 4 μ long and 0.5 μ wide. The organisms may appear individually or in chains. On a Papanicolaou-stained smear, they stain a pale blue. *Bacillus vaginalis* live on the glycogen contained in the vaginal cells, and the optimal environment is at pH 3.9. The glycogen utilized by these lactobacilli is present either in the exfoliated cells or in the vaginal secretion itself. These bacteria sometimes can be associated in the vagina with other organisms such as streptococcus, staphylococcus, enterococcus, diptheroid bacteria, and even colibacillus.

The normal flora (*Bacillus vaginalis*) usually appears during the first week of life. In certain cases, the vaginal flora can be completely absent. The amount of flora fluctuates with the menstrual cycle, as shown in Fig. 8-3.

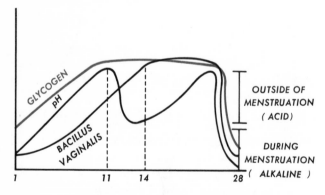

Fig. 8-3. Relationships of vaginal cycles.

Lactic Acid and Vaginal pH

pH is a measure of the degree of acidity or alkalinity. A pH of 7 is neutral, above 7 is alkaline and below 7 is acidic. The vagina usually has an acid pH which varies during the menstrual cycle. The acidity is related to the amount of lactic acid present in the vagina, and the amount of

lactic acid depends on how much glycogen is present. Since the *Bacillus vaginalis* converts the glycogen to lactic acid, the presence of these bacteria enters into this relationship. The amount of glycogen is hormone-dependent; consequently, the degree of acidity or alkalinity (pH) varies during the endocrine cycle (Fig. 8-3).

The Vagina and Sexual Activity

The vagina can respond to sexual stimulation, either physical or psychological, in the following manner. First, fluid filters through the walls of the blood vessels in the vaginal wall and lubricates the vagina. The muscular layer relaxes, dilating the vaginal cavity, and making sexual union possible. During orgasm, the vaginal muscles contract, spasmodically at first, then regularly several times.

Normally, this process is not painful. However, some women are unable to have sexual intercourse because of intense pain. Aside from physical and psychological reasons, over half of these patients suffer from vaginitis, or vaginal infection.

The Vaginal Mucosa

The mucosa, or superficial layer of the vaginal wall, is composed of squamous stratified epithelium, similar to skin, except that it is not keratinized. This layer can be studied with sections and with smears. We will first discuss the cross section of this layer in order to illustrate the structure.

Section of the Vaginal Epithelium

Fig. 8-4 shows a very schematic cross section of the vaginal mucosa. There are three layers in the human adult. The basal layer is the source of the entire epithelium. Cells which develop here are pushed upward by new cells. As they move to the more superficial layers, they undergo certain changes. The middle layer or midzone, is the thickest part of the epithelium. The superficial layer is the layer which sheds most of the cells seen in a vaginal smear.

Fig. 8-5 shows a more detailed picture of the basal layer, or basal lamina. It is also called the stratum germinativum and it gives rise to the rest of the cells. This germinative zone is formed from one or two layers of small cells, each cell having a large nucleus. These cells, located on the

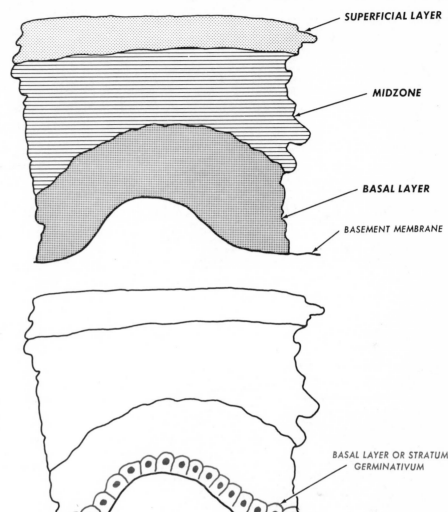

Fig. 8-4. Cross section of vaginal epithelium.

SUPERFICIAL LAYER

MIDZONE

BASAL LAYER

BASEMENT MEMBRANE

Fig. 8-5. Basal layer of vaginal epithelium.

BASAL LAYER OR STRATUM GERMINATIVUM

basement membrane, have a basophilic cytoplasm. (Basophilic, or cyanophilic, means the substance being stained "likes" basic stains, that is, it stains blue.) These cells are the predecessors of the cells found higher up in the section, which are merely phases of maturation of the basal cells. The basophilic cytoplasm in the basal cells will change as the cells mature.

Around the basal cells, there are several layers of parabasal cells, which are larger than the basal cells and are linked together with intercellular bridges. The cytoplasm is still basophilic and the nuclei are still relatively large (Fig. 8-6).

The intermediate layer is made up of intermediate cells; that is, cells which are a step between the less mature basal and parabasal cells and the mature superficial cells (Fig. 8-7). These cells, which are attached to one another by intercellular bridges, are numerous and not as round as the

two deeper layers. They have a flattened polygonal appearance and are larger than the basal and parabasal cells. The nuclei of the intermediate cells are smaller than those of the parabasal cells, and the cytoplasm is basophilic. If, for hormonal reasons, maturation of the epithelium is arrested, these intermediate cells may form the superficial layer.

Fig. 8-8 shows the addition of the superficial layer. It is made up of several layers of mature cells. These mature cells are acidophilic, are more loosely held together, and have no intercellular bridges. The superficial cells are larger than the intermediate cells and their nuclei are smaller and pyknotic. No further growth is possible after this stage, and the presence of pyknotic nuclei shows that mitosis is no longer possible either. These superficial cells of the vagina do not normally form keratin. Since new cells are pushing

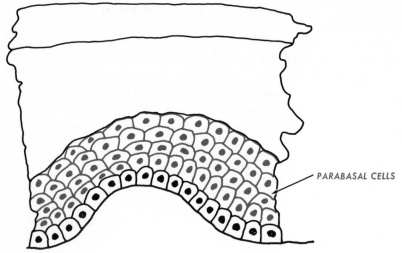

PARABASAL CELLS

Fig. 8-6. Parabasal layer of vaginal epithelium.

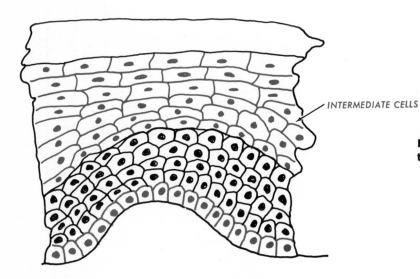

INTERMEDIATE CELLS

Fig. 8-7. Intermediate layer of vaginal epithelium.

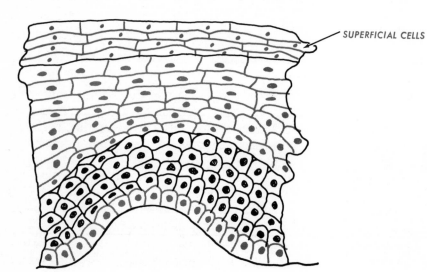

SUPERFICIAL CELLS

Fig. 8-8. Superficial layer of vaginal epithelium.

their way up at a steady rate, and since these cells are the last step in the cell line and are loosely held together, they are regularly shed or exfoliated. In general, as these cells mature, they flatten and get larger, while the nuclei become smaller, and they lose their intercellular bridges as superficial cells (Fig. 8-9). The whole maturation process takes several days, but it has been shown that the administration of estrogen can double this rate.

Fig. 8-11 shows that about the fourteenth or fifteenth day, the epithelium has reached its full development, and is 230 to 300 μ thick. This is about the time of ovulation and the end of the estrogenic phase.

After ovulation, the luteal or desquamation phase begins. The superficial cells swell, causing the epithelium to appear thicker. The epithelium reaches its maximum thickness near the seventeenth day. These cells usually desquamate, or fall

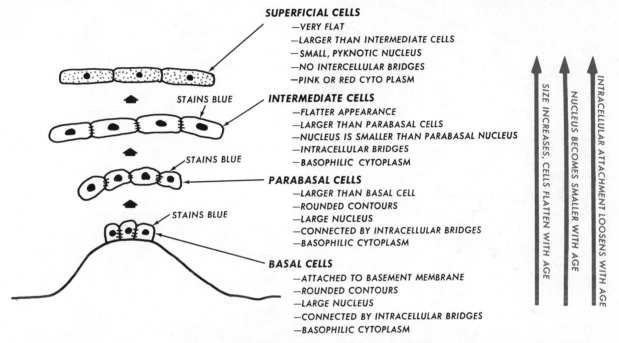

SUPERFICIAL CELLS
— VERY FLAT
— LARGER THAN INTERMEDIATE CELLS
— SMALL, PYKNOTIC NUCLEUS
— NO INTERCELLULAR BRIDGES
— PINK OR RED CYTO PLASM

STAINS BLUE

INTERMEDIATE CELLS
— FLATTER APPEARANCE
— LARGER THAN PARABASAL CELLS
— NUCLEUS IS SMALLER THAN PARABASAL NUCLEUS
— INTRACELLULAR BRIDGES
— BASOPHILIC CYTOPLASM

STAINS BLUE

PARABASAL CELLS
— LARGER THAN BASAL CELL
— ROUNDED CONTOURS
— LARGE NUCLEUS
— CONNECTED BY INTRACELLULAR BRIDGES
— BASOPHILIC CYTOPLASM

STAINS BLUE

BASAL CELLS
— ATTACHED TO BASEMENT MEMBRANE
— ROUNDED CONTOURS
— LARGE NUCLEUS
— CONNECTED BY INTRACELLULAR BRIDGES
— BASOPHILIC CYTOPLASM

SIZE INCREASES, CELLS FLATTEN WITH AGE

NUCLEUS BECOMES SMALLER WITH AGE

INTRACELLULAR ATTACHMENT LOOSENS WITH AGE

Fig. 8-9. Maturation changes in the vaginal epithelium.

The Vaginal Epithelium Cycle

Like the rest of the female genital tract during the childbearing years, the vaginal epithelium undergoes cyclic changes under the influence of the ovulatory cycle (Fig. 8-10). There are two main phases: the proliferative, or estrogenic phase; and the desquamative or luteal phase. Fig. 8-11 shows these cyclic changes. Note the epithelium at the beginning of its cycle, that is, during menstruation. At this point, the epithelium is as little as 150 to 180 μ thick. The basal layers are undergoing intense mitotic activity and producing many rounded cells. Later, the parabasal and intermediate layers increase in thickness. The superficial layers increase as the lower layers mature, and the superficial cells will appear eosinophilic. Eosinophilic cells are readily identified because they have an affinity for red colors in histological stains.

off. At the same time, intracellular glycogen decreases. This is the end of the cycle, also known as the premenstrual phase of the vaginal epithelial cycle. The area is infiltrated by white blood cells or leukocytes, and is only 150 to 180 μ thick. The luteal phase is over and the estrogenic phase is ready to begin.

It is important to bear in mind that changes in the histology of the vagina are not as clear-cut and obvious as those which occur in such other regions as the endometrium. Changes vary with the region of the vagina. For example, the changes are most clear in the upper third of the lateral vaginal wall.

The Vaginal Smear

Since it would be impractical and even dangerous to remove a slice of tissue from the vaginal wall every time an examination was necessary,

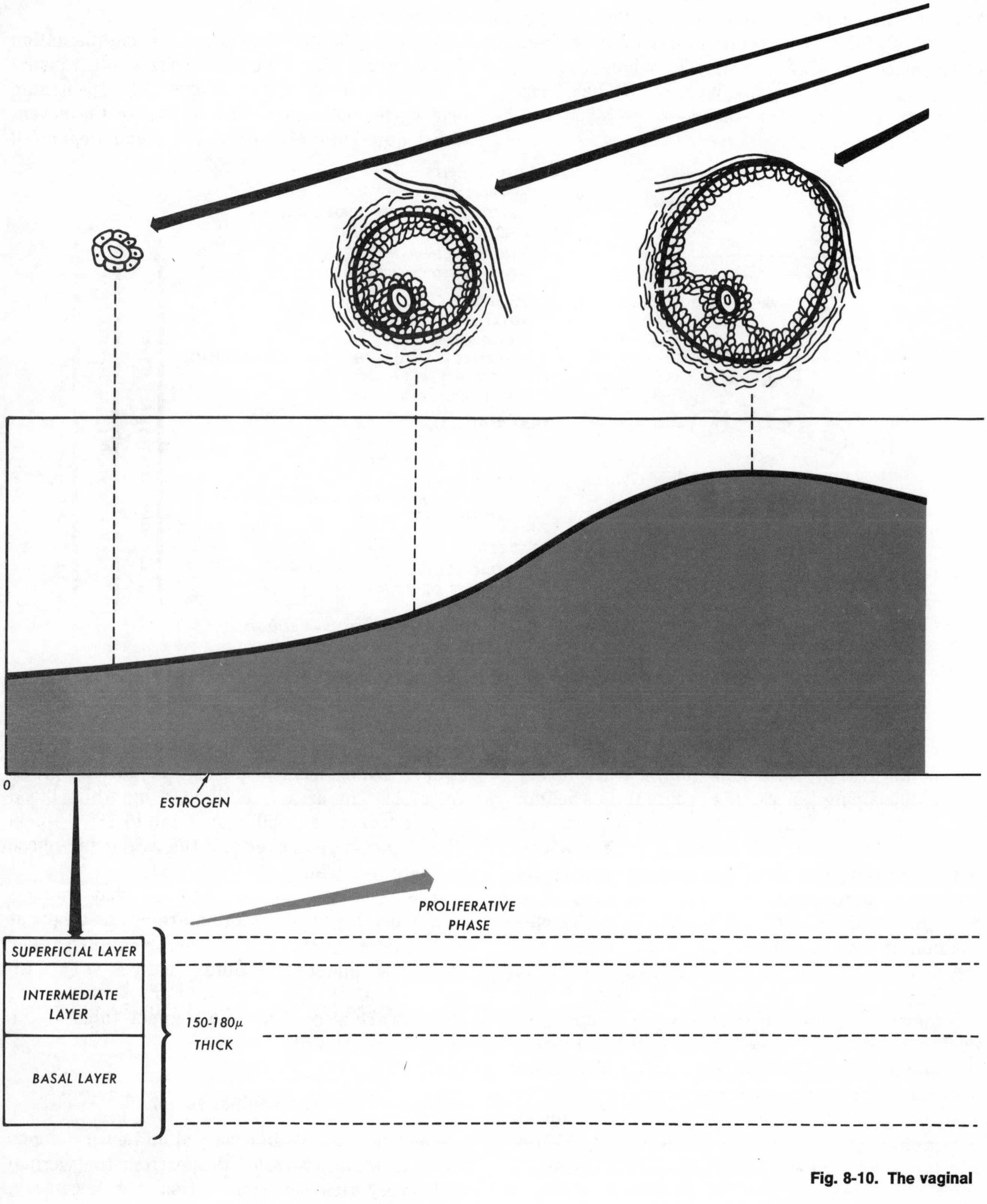

ESTROGEN

PROLIFERATIVE
PHASE

SUPERFICIAL LAYER

INTERMEDIATE
LAYER

150-180μ
THICK

BASAL LAYER

Fig. 8-10. The vaginal

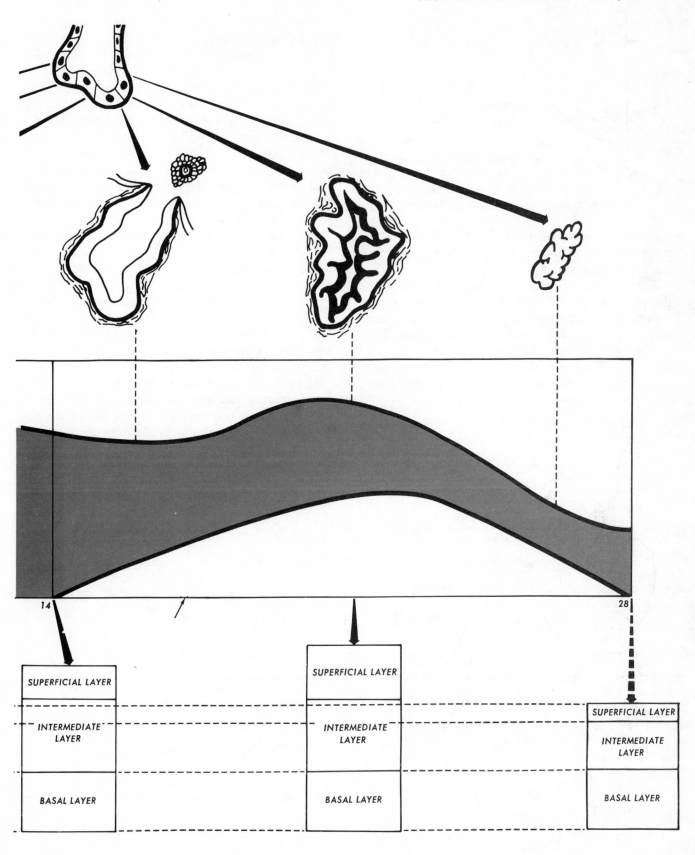

SUPERFICIAL LAYER

INTERMEDIATE LAYER

BASAL LAYER

SUPERFICIAL LAYER

INTERMEDIATE LAYER

BASAL LAYER

SUPERFICIAL LAYER

INTERMEDIATE LAYER

BASAL LAYER

epithelial cycle.

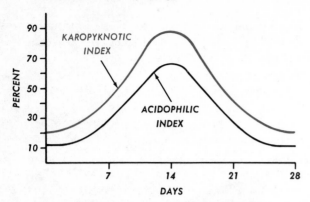

Fig. 8-11. Cyclic changes in vaginal epithelium.

the vaginal smear is used. Most of the cells in the smear normally originate from the superficial or intermediate layers.

For the sake of illustration, we will begin with the most superficial cells. Fig. 8-12 shows some superficial squamous cells. Note that they are large polygonal cells, with flat cytoplasm. The nuclei are pyknotic, that is shrunken and condensed, indicating that these cells are no longer viable. Some demonstrate karyorrhexis or frag-

mented pyknotic nuclei. These cells are 40 to 50 μ in size, on the average, but variations are not uncommon. Using Papanicolaou stain to color these cells, the cytoplasm takes on a pink tint because of its eosinophilic (or acidophilic) properties. (When preparing these smears, beware of drying out, since dryness increases the eosinophilic color.) Sometimes the cells will stain a light blue (basophilic). Remember that the most mature cells in this cell line are eosinophilic. Sometimes granules are present in the cytoplasm, either around the nucleus (perinuclear) or in the peripheral portions of the cell.

Complete maturation of the epithelium requires the presence of estrogen, so a good indication of adequate estrogenic activity is the presence of pyknotic nuclei in the superficial cells. If, for some reason, full maturity cannot be reached, the intermediate cells form the majority in the smear.

Fig. 8-13 shows some examples of intermediate squamous cells. They are either the same size or smaller than the superficial cells, with a basophilic cytoplasm that stains more densely than the more mature forms. In some cells, however, eosinophilia is possible. The main difference between the superficial and the intermediate cells is the nuclei. The intermediate cells have a rounded or oval nucleus, with a well-preserved nucleoplasm and nuclear membrane. The nuclei show chromatin granules and chromocenters. This type

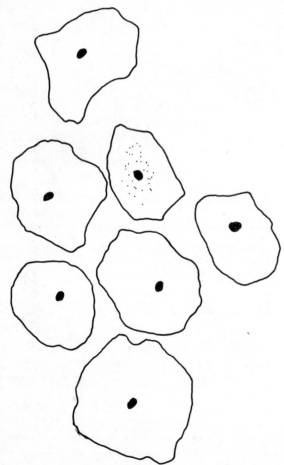

Fig. 8-12. Superficial squamous cells.

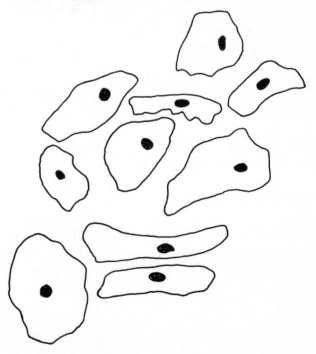

Fig. 8-13. Intermediate squamous cells.

of nucleus is sometimes referred to as a vesicular nucleus. In this category are also included some cells with slightly degenerated nuclei. A common variant is the navicular (boat-shaped) cell. These cells contain glycogen and are frequently seen in pregnancy and menopause.

Fig. 8-14 shows some parabasal cells. In the normal smear taken from women of childbearing age, these cells are not as common as the intermediate and the superficial cells. These cells can increase in number in cervical pathology. They are smaller than the more superficial cells, meas-

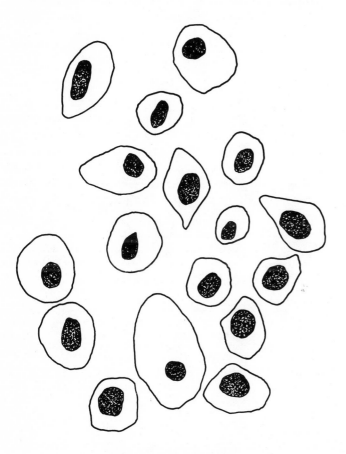

Fig. 8-14. Parabasal cells.

uring about 12 to 30 μ and have a basophilic cytoplasm which sometimes contains small vacuoles. The cells have a round or oval shape and smooth cytoplasmic borders due to contraction of the cytoplasm after death. Cells rounded in this way are normally cast off by the vaginal wall, whereas sometimes they are seen in clumps or having polygonal sides, which means they were scraped off by force during the removal of the smear specimen. These cells have oval or round nuclei of the same size as the intermediate cell nuclei and

possess the same vesicular nucleoplasm (chromo-centers and chromatin). Although the nuclei are the same size as those of the intermediate cells, they have less cytoplasm.

Because they are located so deeply in the epithelium, basal cells are rarely seen in smears. Their presence may indicate that a pathologic process has damaged the more superficial layers. These cells appear as very small parabasal cells, with a scanty basophilic cytoplasm. The nucleus is relatively large and has chromatin granules.

Cyclic Variations in the Smear

Figs. 8-15 and 8-16 are illustrations of smears taken at intervals during the menstrual cycle. Various methods have been devised to measure changes in the smear due to changes in the endocrine cycle. One of the methods is known as the acidophilic index. This index is the measure of the percentage of the superficial squamous epithelial cells that are eosinophilic. Fig. 8-16 shows that eosinophilia in these cells reaches a peak at ovulation. This peak can be as high as 50% to 75% of all the squamous superficial cells counted.

The karyopyknotic index is the measure of the percentage of the squamous superficial epithelial cells that have pyknotic nuclei. This index also shows a peak at ovulation and can reach 50 to 85%. Bear in mind that much variation is possible from patient to patient. At ovulation, the per-

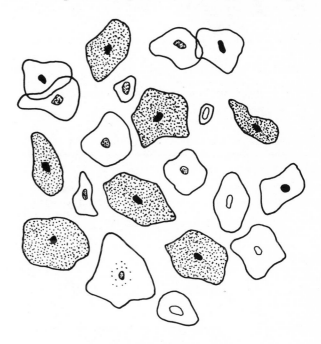

Fig. 8-15. Preovulation smear.

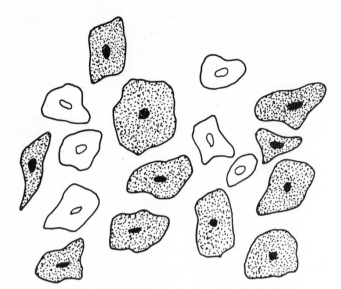

Fig. 8-16. Smear right after ovulation.

Endocervical cells usually are not seen in vaginal smears, but are visible on smears taken from the endocervix. These cells are seen as well-preserved cells with a basal nucleus and a columnar shape. Their cytoplasm is slightly basophilic and the cells are often seen in clusters. At the peak of endocrine activity, "nipples" appear at the ends of these cells, indicating increased secretory activity.

centage of the different cell types on a normal smear is as follows: 0% parabasal, 40% intermediate, 60% superficial.

The Endocervix

Fig. 8-17 shows the location of the endocervix. Histologically, it has a different structure from the portio and the change from endocervix to portio occurs abruptly at the squamocolumnar junction. The endocervix is lined with a single layer of columnar epithelial cells, which suddenly stop and become the stratified squamous epithelium of the portio vaginalis.

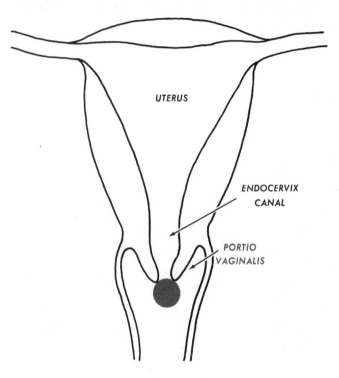

Fig. 8-17. The cervical canal.

Review Questions

1. In the illustration below, label: Basement membrane, basal cells, parabasal cells, intermediate cells, superficial cells.

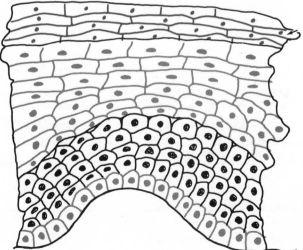

2. Match with illustrations below and on p. 82:
   large flat polygonal cells _____
   small, rounded cells _____
   large polygonal cells with pyknotic nuclei _____
   large polygonal cells with rounded nuclei larger than above _____
   most mature cells _____
   not present in the absence of estrogen _____
   small, rounded cells with reduced cytoplasm _____

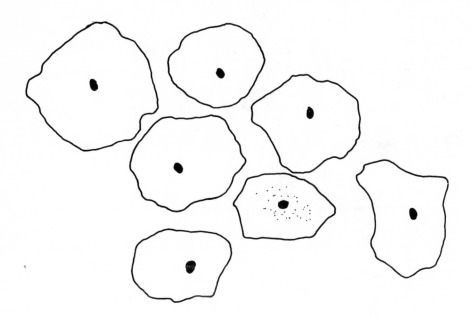

Superficial squamous cells.

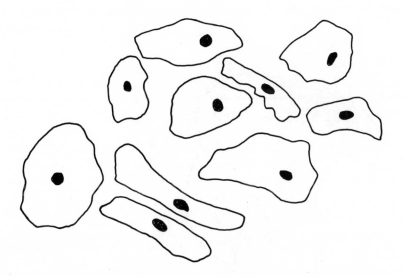

Intermediate squamous cells.

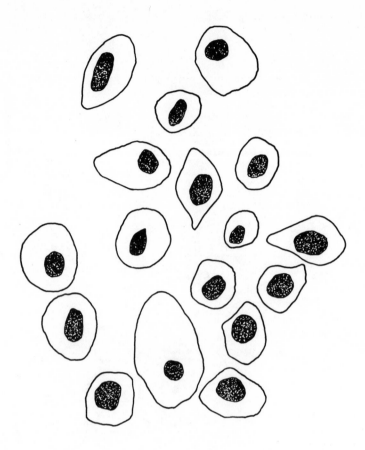

Parabasal cells.

chapter 9

The Urinary Tract

The body produces numerous waste products from its many metabolic processes. Some come directly from energy production, others are by-products of the breakdown of various substances, such as blood cells. These waste products are present in circulating blood and, if not removed, will accumulate and become toxic. The kidneys play a major role in the removal of these wastes, excreting them as urine. The kidneys, as we will see, also are important in regulating the pH and osmotic balance of the body.

Fig. 9-1 represents the main components of the urinary tract. Urine, produced by the kidneys, passes through the ureters into the bladder. When a certain quantity accumulates, the bladder releases the urine into the urethra, which is a passage to the outside.

The Kidneys

Fig. 9-2 is a simplified illustration of a kidney. Urine is produced in the medullary region of the kidney and passes into the calices (singular; calyx), which form the pelvis of the kidney. From the pelvis, urine passes into a ureter, then into the bladder.

Fig. 9-3 shows a blowup of a nephron, a long tube lined with an epithelium. There are over a million nephrons in each kidney, each nephron a functional unit which produces urine. Each nephron begins at the renal corpuscle and winds its way to the collecting duct. Blood vessels enter and leave the renal corpuscle, but only urine passes through the nephron to the collecting duct.

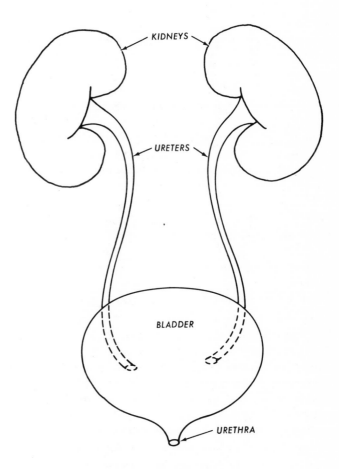

Fig. 9-1. The urinary tract.

The urine from the collecting ducts empties into a calyx, passes through the pelvis and finally into the ureter.

Fig. 9-4 highlights the renal corpuscle. The epithelial cells form the covering of the corpuscle,

83

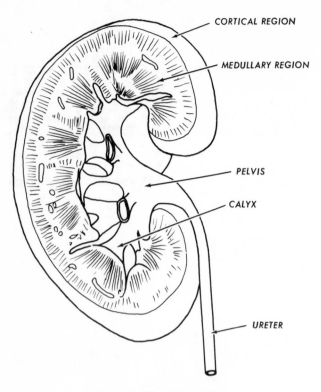

CORTICAL REGION

MEDULLARY REGION

PELVIS

CALYX

URETER

Fig. 9-2. Simplified kidney.

known as Bowman's capsule. The glomerulus is defined as the complex capillary network inside this capsule. Blood enters through an arteriole and is directed through these capillaries, which then empty into another arteriole (not a venule) which carries the "purified" blood back into the circulation. The capillary network shown in Fig. 9-4 is representative of the actual glomerulus, which is a ball of vessels winding around each other.

Fig. 9-5 represents the glomerular filtration process, a forced process due to blood pressure. If blood pressure in these vessels is too low, glomerular filtration cannot occur and serious consequences may result. As you know, blood contains solids as well as liquids. Red and white blood cells make up almost half the volume of the fluid we call blood. The remainder is plasma, a protein solution which contains water as well as hundreds of different chemical substances. Since it would not be practical to remove blood cells or proteins from the circulating blood, the glomerulus is selectively permeable. This dialysis process occurs at the level of the capillary walls. The substances allowed to leave the capillary, known as the filtrate, must pass across pores in the epithelial cells and through the basement membrane of the capillary. Molecules larger than a molecular

weight of about 68,000 cannot pass across this barrier, and, since most blood proteins have a larger molecular weight than this, very little protein is found in the urine.

Blood flows quickly through the body, the same blood passing through the kidneys every five minutes. If you were to calculate how much filtrate is produced, you would find that the glomerulus produces almost 200 liters of filtrate every 24 hours. Obviously, it would be impossible to excrete so many times your own weight in fluids each day, so it seems logical that about 99% of this filtrate is reabsorbed by the nephron. The blood passes through the glomerulus, forming a glomerular filtrate, which contains water and dissolved glucose, urea, uric acid, phosphates, creatinine, electrolytes, amino acids, and other substances. Although this filtrate is not yet urine, we will refer to it as glomerular urine and follow its progress through the nephron and indicate the changes that take place as it gradually becomes urine.

The cells of the nephron, although all epithelial cells, vary with the segment. Each segment has its own specific function of absorption (and sometimes, excretion). For the sake of simplicity, we will consider the tube as a whole. It is important to bear in mind that water, as in the cell itself, passively follows the osmotic changes which take place as each substance is reabsorbed. Of course, the only substances reabsorbed are those which are useful to the body, while the waste products become more and more concentrated as the urine nears the end of the nephron.

An important function of the kidney, aside from waste removal, is the role it plays in keeping the blood pH stable. Since the kidney can selectively secrete acid or base substances, it can maintain a steady state.

Earlier, we mentioned the importance of sodium and potassium concentrations in cellular function. If the plasma sodium and potassium balance is upset, the balance of the cells also will be affected. The kidney acts to keep these electrolytes balanced. These functions have been simplified and are actually much more complex than we have described here. Some of them, such as electrolyte balance, are controlled by the endocrine system.

In summary, then, we can say that the main function of the kidneys is to remove wastes from the blood, as well as help maintain the proper acid-base balance as well as the electrolyte balance of the body.

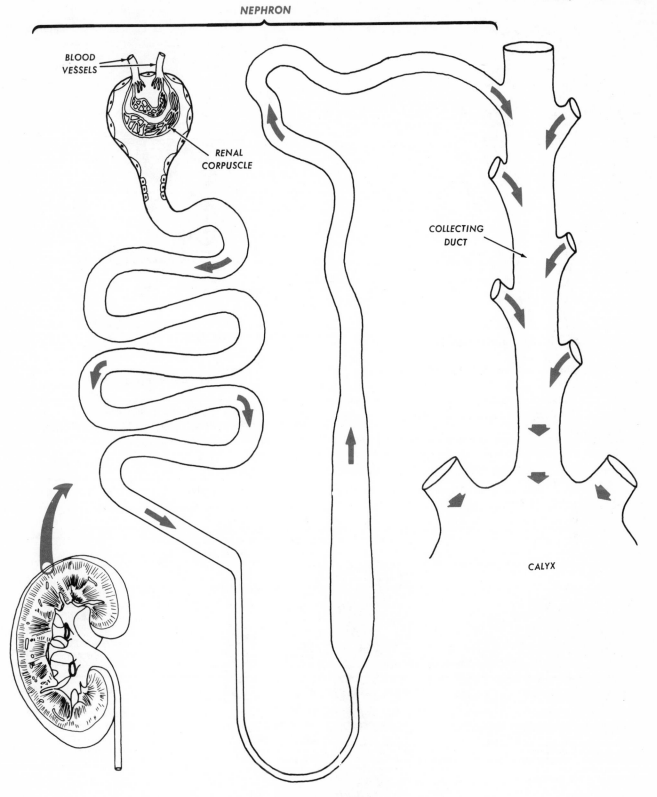

NEPHRON

BLOOD
VESSELS

RENAL
CORPUSCLE

COLLECTING
DUCT

CALYX

Fig. 9-3. Nephron.

The Excretory Passages

Urine produced by the kidneys does not merely flow downhill to escape the body—it is pushed along by the excretory passages. Each part of the tract helps push urine along by contracting in sections, so the urine is passed along in waves into the bladder. When the bladder is full, it also con-

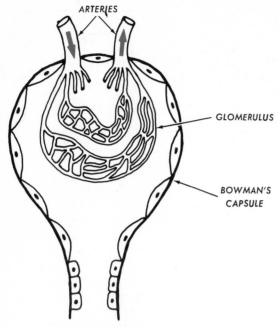

Fig. 9-4. Renal corpuscle.

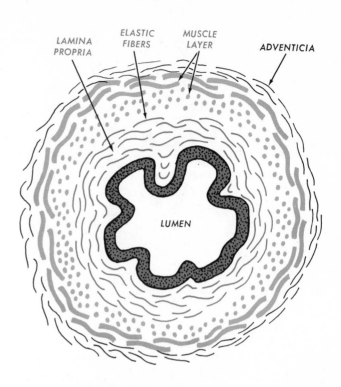

Fig. 9-6. Cross section of ureter.

tracts and pushes the accumulated urine out of the body.

The Renal Pelvis and Ureters

The pelvis and ureter are continuous with each other and have a similar epithelium. What we describe for the ureter epithelium is also true for the pelvis. Fig. 9-6 shows that the lumen, or opening of the tube, is irregularly shaped and surrounded by a thick lamina propria which contains many elastic fibers. Muscle fibers surrounding the tube are covered by the elastic adventicia. The ureter is very elastic, and it is capable of distension as well as contraction to move urine down to the bladder in pulses. Fig. 9-7 shows the epithelium, known as the transitional epithelium or

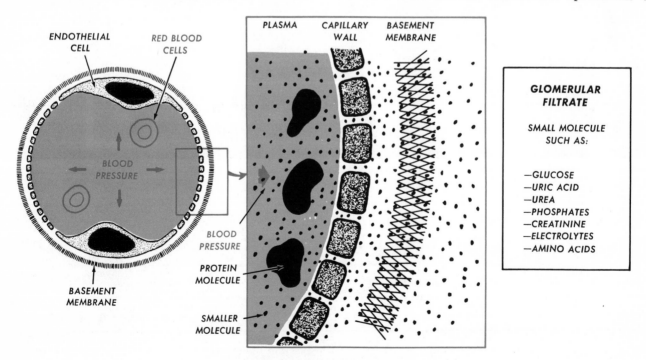

Fig. 9-5. Glomerular filtration process.

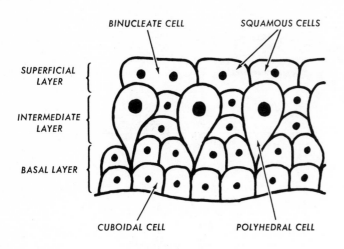

Fig. 9-7. **Transitional epithelium (not stretched).**

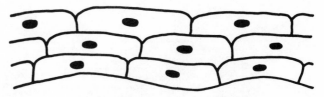

Fig. 9-8. **Transitional epithelium (stretched).**

urothelium. This type of epithelium is found from the pelvis to the urethra and can change from columnar shape to squamous shape, depending upon how much it is stretched. The basal layer, on a basement membrane, consists of high cuboidal cells. The intermediate layer is made up of cuboidal and polyhedral cells (which look like tennis rackets), and the superficial layer can be either squamous or cuboidal cells. Note that some of the superficial cells are binucleate, that is, they have two nuclei. Fig. 9-8 shows the same section of the epithelium being stretched. It has become much thinner, and most of the cells appear flattened and seem to incorporate keratin which makes them less permeable to water.

The Bladder

Fig. 9-9 shows the wall of the bladder. Note that the epithelial layer is actually folded to allow for stretching and shrinking. The lamina propria layer is very rich in elastic fibers and contains nerve and areas of mucus-secreting columnar cells. The muscular layer, sometimes referred to as the detrusor muscle, is well developed and is responsible for contracting to empty the bladder. Another muscular structure (not shown in the figure) is known as the sphincter. Its function is to keep the neck of the bladder closed except during urination. The epithelium of the bladder looks similar to that of the pelvis and ureter, but is thicker.

The Urethra

Since the male urethra is quite different from the female urethra, we will discuss the latter here and the male urethra in a later section. The female urethra is lined with transitional epithelium in the areas closest to the bladder. As we move away

Fig. 9-9. **Wall of the bladder.**

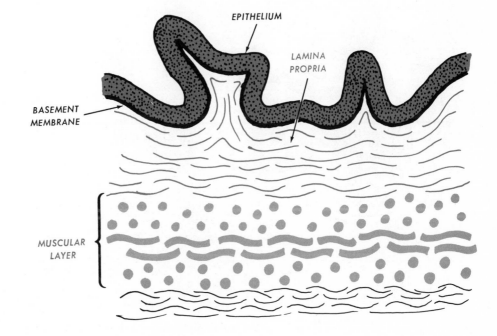

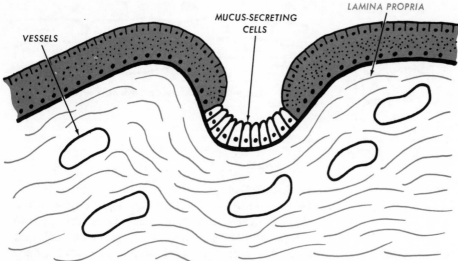

Fig. 9-10. Female urethra.

from the bladder, the epithelium changes to stratified squamous, which then becomes columnar. At
the end of the urethra, where it opens to the exterior, the epithelium appears as epidermoid epithelium. In certain places, the epithelium forms
depressions which contain mucus-secreting cells
(Fig. 9-10). The lamina propria layer consists of
thick, elastic connective tissue and contains numerous blood vessels (Fig. 9-11). Note also the
muscular layer, which appears well developed.

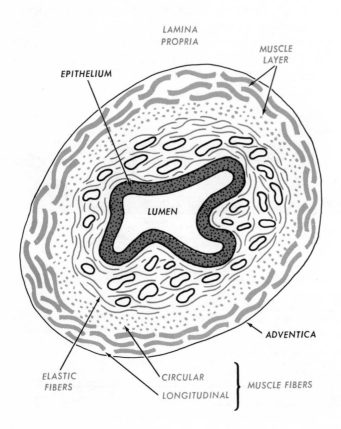

Fig. 9-11. Female urethra.

Review Questions

1. For the illustration below:
    Title _____
    Label: kidneys, uterers, urethra, bladder

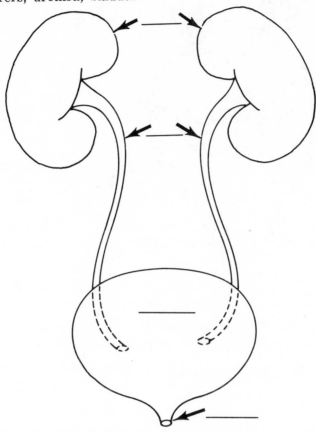

2. For the illustration below:
    Title _____
    Label: calyx, ureter, pelvis, cortical region, medullary region.

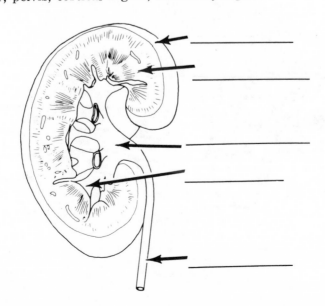

3. For the illustration below:

   Title _____

   Label: Bowman's capsule, arteries, glomerulus

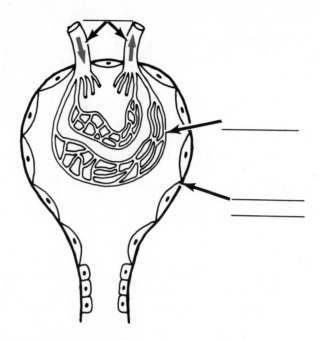

4. Match the following with definition opposite:

   Plasma _____
   Filtrate _____
   Dialysis _____

   a) passage across a semipermeable membrane
   b) substance leaving any capillary
   c) liquid portion of blood

5. for the illustrations below:

   Title _____ Title _____

   Label: squamous cells, cuboidal cell, polyhedral
   cell, binucleate cell, superficial layer, basal
   layer, intermediate layer.

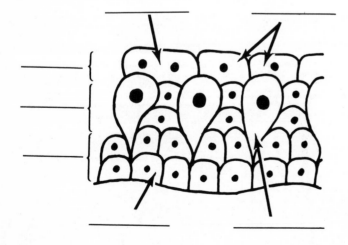
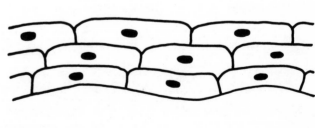

chapter

10

The Male Genital Tract

Fig. 10-1 gives an overall view of the male genital tract. Urine from the bladder passes through the urethra to the exterior, sharing the same passage to the outside as the semen released during ejaculation.

Semen consists of both the male sex cells known as spermatozoa (produced in the testis) and the fluids secreted by the various glands along the tract. The seminal vesicles, Cowpers glands and the tubes of the whole tract contribute to the seminal fluid. During sexual activity, prior to ejaculation, semen is stored briefly in the prostate until violent muscular spasms propel the liquid through the urethra and outside the body. There are mus-

cular structures which close off the bladder at these moments and others which propel the semen.

The testicles, located in the scrotum (Fig. 10-1), are the source of the sperm cells. They are similar to the ovaries in that they too have a double function: they are the seat of both endocrine and exocrine activity. The exocrine role is the production of sperm cells. Unlike the female, who is born with a lifetime supply of sex cells, the male is constantly producing sperm cells. These cells are produced in the seminiferous tubules (Fig. 10-2), and eventually end up in the epididymis, a winding tube 5 to 7 meters or about 20 ft. long. Fig. 10-1 gives an overall picture of the

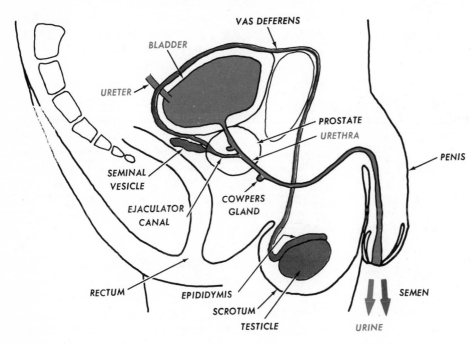

Fig. 10-1. The male genital tract showing the path taken by urine, and the path taken by semen.

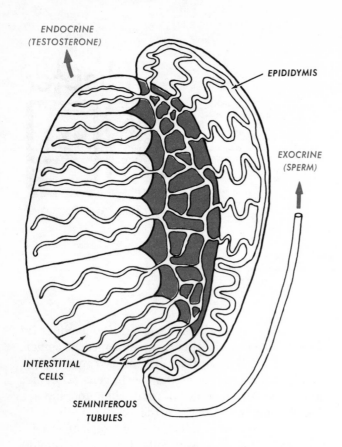

Fig. 10-2. The testicle.

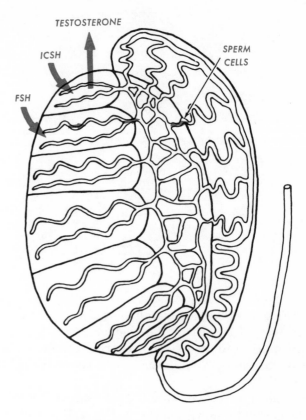

Fig. 10-3. The endocrine testicle.

path followed by the sperm cells: from the epididymis, they pass into the ductus deferens (or vas deferens), then through the ejaculator canal into the prostate gland. Inside the prostate, the ejaculator duct empties into the urethra and passes through the penis to the outside.

The endocrine testicle (Fig. 10-3), unlike its female counterpart the follicle, constantly secretes the male hormone, testosterone, in amounts which vary from time to time. The male hormone comes from the interstitial cells, located between the seminiferous tubules of the testicle. Testosterone is responsible for the secondary sex characteristics, such as deep voice, and body hair distribution. As the ovary in the female is controlled by the hypothalamus and pituitary, so the testicle is under a similar influence. FSH stimulates the development of sperm cells, while the equivalent of LH, known as ICSH (interstitial cell stimulating hormone) stimulates the interstitial cells (Fig. 10-3).

The epididymis plays several different roles. As the sperm are moved along this passage, they are subject to the secretions of its wall. These secretions give the sperm the ability to move, since they are immobile in the seminiferous tubules, and the ability to fertilize the ovum. The muscular walls of the epididymis are capable of rhythmic contractions, similar to peristalsis (see Fig. 10-4). (Peristalsis is a progressive wavelike movement which occurs in hollow tubes. It is due to the contraction of circular muscles in the walls, which squeeze the contents along the pathway.) The mucosa of the epididymis consists of a pseudostratified epithelium made up of small basal cells and tall columnar cells with cilia (Fig. 10-4).

The ductus deferens or vas deferens (Fig. 10-5) receives the contents of the epididymis and passes it via its thick muscular walls into the body, to a zone of the ductus deferens known as the ampulla (Fig. 10-6). The ampulla, which acts as a reservoir for sperm and the fluid secreted by the ductus deferens, is capable of expansion. It also is capable of peristaltic contractions, and, like the epididymis, has a pseudostratified columnar epithelium.

The ejaculatory duct (Fig. 10-6), receives the contents of the ampulla as well as the secretions of the seminal vesicles. This duct is a simple passage through the prostate, emptying into the urethra. Its epithelium consists of simple columnar or pseudostratified columnar cells. The structures

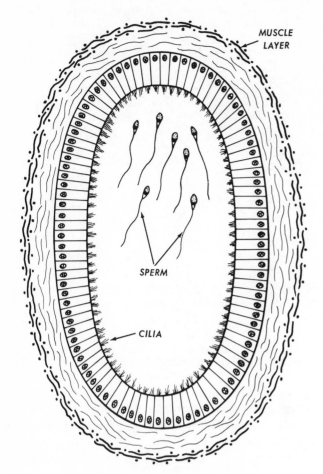

Fig. 10-4. Cross section of epididymis.

The labels on this figure read: MUSCLE LAYER, SPERM, CILIA.

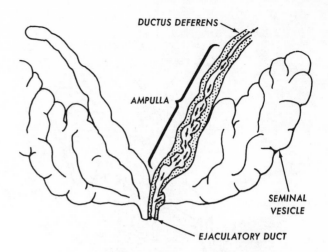

The labels on this figure read: DUCTUS DEFERENS, AMPULLA, SEMINAL VESICLE, EJACULATORY DUCT.

Fig. 10-6. Vas deferens and seminal vesicle.

Aside from the passages mentioned above, there are several accessory glands which add to the seminal fluid. The seminal vesicle (Fig. 10-7) is composed of many tubes which secrete a fluid containing several substances, notably fructose, which is the food utilized by the living sperm cells. The tubes of the seminal vesicle are lined either with a simple columnar or pseudostratified epithelium (Fig. 10-8). These secretions, as mentioned above, empty into the ejaculatory duct. The prostate will be covered in detail later, but, for the moment, consider it as an organ composed of muscular and glandular tissue which produces its own secretions and is capable of contractions which assist the sperm into the urethra and out of the body. The bulbourethral glands of Cowper are pea-sized glands which empty into the urethra. Their secretion is similar to prostatic fluid.

The urethra in the male consists of three segments, as shown in Fig. 10-9. The prostatic urethra consists mainly of a mucosa, since the muscular layer is combined with the fibromuscular tissue of the prostate. The epithelium (Fig. 10-9) is of the transitional type above the level where the ejaculatory ducts enter. It changes to a stratified columnar or pseudostratified epithelium for the rest of the urethra, except at the orifice. The membranous urethra consists of a mucosa and a muscular layer. The mucosa is made up of a stratified columnar epithelium with a basal layer of small cells, two or three layers of polygonal cells, and a superficial layer of columnar cells. The lamina propria is made up of connective and elastic tissue with glands. The muscular layer is more developed near the prostate and decreases lower down. The urethra in the penis is similar to that

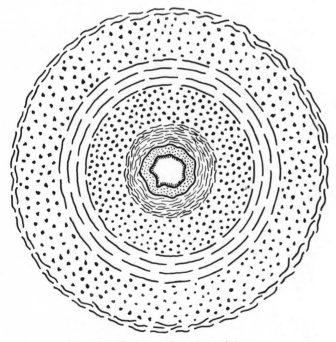

Fig. 10-5. Cross section of vas deferens.

from the testicles to the ejaculatory ducts are bilateral, that is, there are two of each. They come together at the urethra.

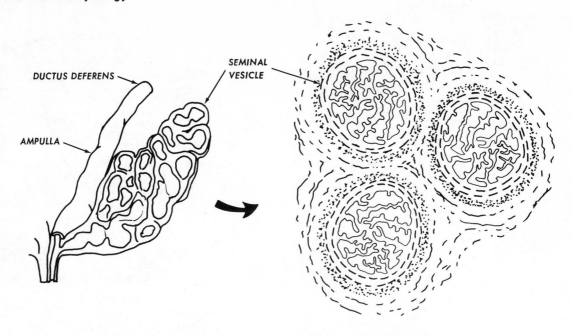

Fig. 10-7. Cross sectional section of the tubules of the seminal vesicle.

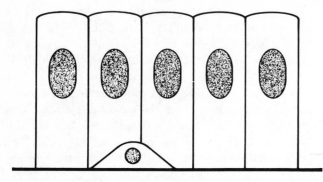

Fig. 10-8. Epithelium of seminal vesicle.

of the membranous urethra, with the exception of the area near the urethral orifice, where it becomes a stratified squamous epithelium like the glans of the penis.

The Prostate

The prostate can be divided into three main layers or zones (Fig. 10-10). The urethra passes through the fibromuscular center, which is surrounded by glandular tissue, which, in turn, is

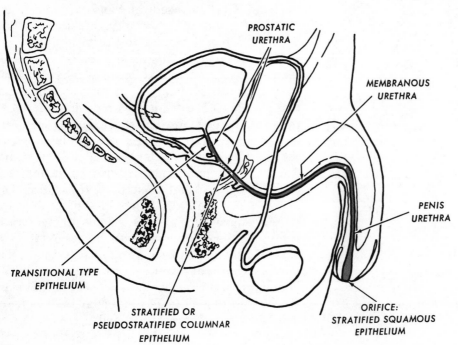

Fig. 10-9. The three segments of the male urethra.

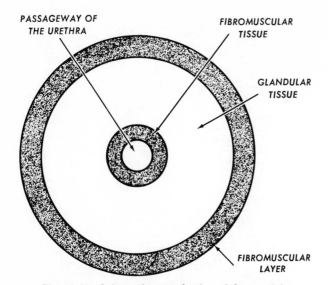

Fig. 10-10. **Schematic organization of the prostate.**

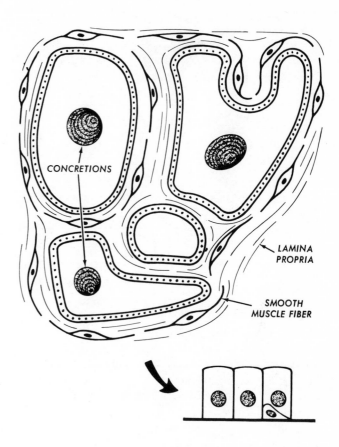

Fig. 10-12. **Cross section of prostate gland.**

surrounded by an elastic fibromuscular layer.

The glandular portion of the prostate is made up of ramified tubes (Fig. 10-11). Fig. 10-12 shows a cross section of several of these tubes. Note that the simple cubic or columnar epithelium is on a basement membrane, which is surrounded by a lamina propria containing connective tissue, elastic fibers, and, most importantly, muscle fibers. These fibers play a role in the sudden spasmodic prostatic secretion. Inside the lumen of these glands, concretions formed by the hardening of the prostatic fluid are visible. These concretions can cause pathological conditions.

The prostatic glands secrete a fluid which supplies the sperm cells with such substances as amino acids, enzymes, and citric acid, which are essential for their survival. These glands are un-

Fig. 10-11. **Simplied representation of prostate.**

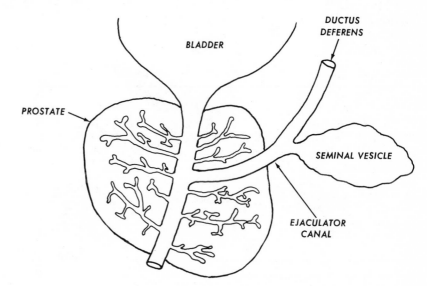

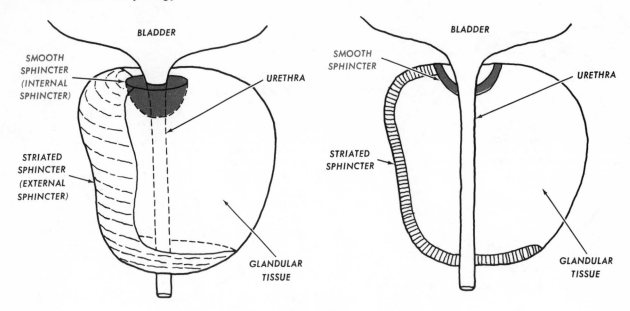

Fig. 10-13. Schematic of prostate gland, and simplified side view.

der the influence of male hormones and they undergo involution if the subject is castrated. Fig. 10-13 shows that the prostate contains two main muscular structures, the smooth sphincter, and the striated sphincter. The smooth, or internal, sphincter of the prostate surrounds the prostatic urethra. It actually corresponds to the muscular layer of this segment of urethra. This smooth sphincter keeps the bladder sealed except during urination. The striated, or external, sphincter of the prostate surrounds the lower end of the prostatic urethra like a muscular layer, but only covers the front of the glandular tissue between the two extremities of the prostatic urethra. This sphincter, which also seals off the bladder, plays an important role in ejaculation by squeezing the prostatic glands into the urethra.

Ejaculation

The process of ejaculation occurs in two stages: secretion of substances into spermatic canals, and expulsion of these substances outside the body.

Prior to ejaculation, the testicle, epididymis, and the ductus deferens produce a steady secretion which accumulates in the ampulla (Fig. 10-14). This secretion is brought to the ampulla by peristaltic contractions of the muscle layers of these structures. The prostate and the seminal vesicles, on the other hand, secrete only at the moment of ejaculation. As these secretion products accumulate in the prostatic urethra in the last moments before ejaculation, their pressure builds up, held back by the striated sphincter below (Fig. 10-15). The smooth sphincter prevents

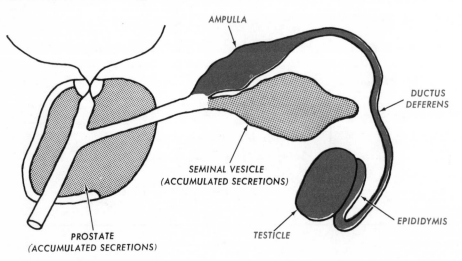

Fig. 10-14. Ejaculatory apparatus in state of secretion.

the semen from being forced up into the bladder, and the ampulla contracts to keep pushing the fluid into the prostatic urethra.

At the moment of ejaculation (Fig. 10-16), the urethra strongly contracts forcing the liquid past the striated sphincter and through the prostate in pulses or waves of contraction. Simultaneously, the prostate and seminal vesicles release their stored secretions and the totality of the semen is forced through the urethra and out of the body.

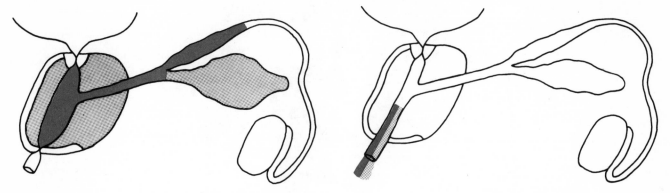

Fig. 10-15. Ejaculatory apparatus expulsion stage.

Fig. 10-16. Moment of expulsion.

Review Questions

1. Label the parts shown in the illustration below.

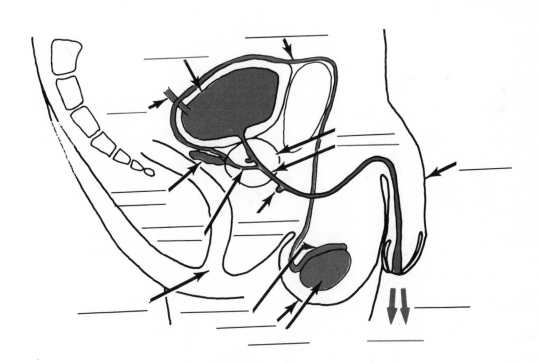

2. Label the parts shown in illustration below.

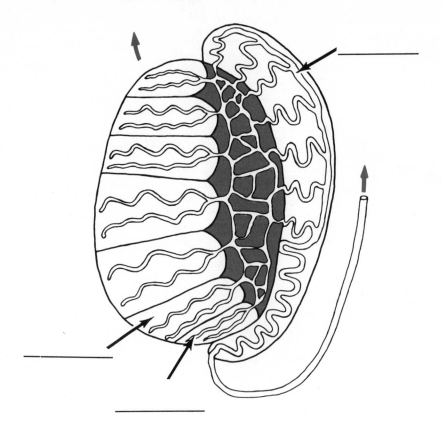

3. Label the parts shown in the illustration below.

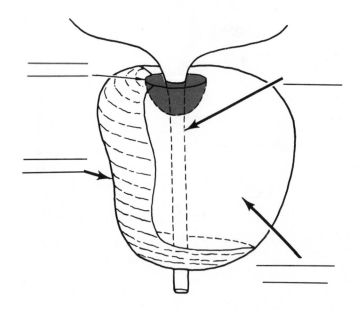

4. Label the following parts shown in the illustration below: vas deferens, seminal vesicle, ampulla, ejaculatory duct

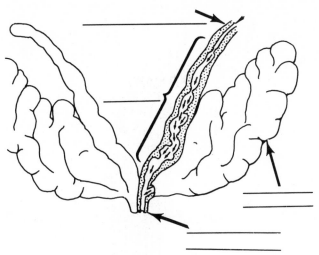

5. Match the following labels with the illustrations below:
   1) vas deferens
   2) prostate gland
   3) epididymis
   4) seminal vesicle

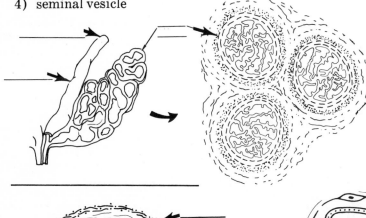

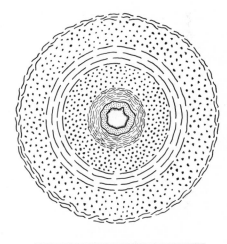

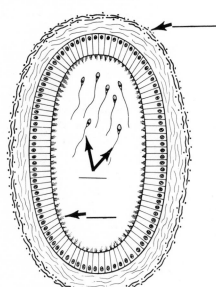

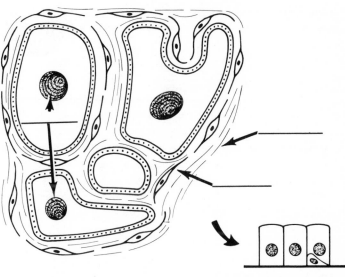

chapter 11

The Respiratory Tract

The living cell requires a constant supply of oxygen as well as a way to remove carbon dioxide. Oxygen is carried by the red blood cells, while carbon dioxide is dissolved in the blood. Therefore, there must be a system by which exchanges between the blood and the air we breathe can take place. This system is known as the respiratory system. Fresh air is brought into the body through a series of tubes, and exchanges between air and blood take place in the lungs.

When you breathe in, or inspire, the diaphragm moves down and certain chest muscles contract to lift the rib cage. This active movement causes the lungs to expand and air is pulled in. Breathing out, or expiration, is a passive act. The lungs are elastic and when the inspiration muscles relax, the lungs shrink back to their normal size, pulling the chest in with them and forcing the air out of the body.

Structure

Fig. 11-1 shows the respiratory tract. Air is taken in either through the nose or the mouth and passes down through the pharynx, larynx, trachea and bronchi (singular, bronchus) which eventually leads to the lungs. The lungs are not simply bags filled with air, but are spongy organs made up of millions of pockets where air and blood can exchange gases.

Air can enter either through the mouth or the nasal cavity. The nasal cavity, because of its structure, serves both to warm the air as it enters the body and to add moisture so the air does not dry the respiratory tract. This region also helps filter out larger dust particles. Air entering through either the nose or the mouth passes through the pharynx.

The pharynx is equipped with a type of "lid" which can close off the airway passages when food is being swallowed, to prevent it from entering the larynx. This lid is known as the epiglottis (Fig. 11-1).

The larynx contains the vocal cords. Air passing over these structures creates voice sounds. The larynx leads to the trachea, or windpipe.

The trachea is a tube composed of rings of cartilage and muscle and lined with a mucous membrane. The lower part of the trachea is divided into two tubes known as the bronchi.

Each bronchus enters a lung, and divides into many branches which then further divide, until each branch ends up at a "dead end" and forms clusters of small sacs known as the alveoli (Fig. 11-2).

An alveolus is a thin-walled sac where air can deliver oxygen to the blood and take away carbon dioxide across a very thin membrane. Blood enters the lungs through the pulmonary artery which divides as often as the bronchi do (Fig. 11-3). The final destination of these branchings and sub-branchings of the artery is the fine capillary network which is found in the walls of the alveoli. The oxygenated blood leaves the capillary and enters the venous network which eventually converges and forms the pulmonary vein which goes back to the heart where it is pumped out to all the cells of the body.

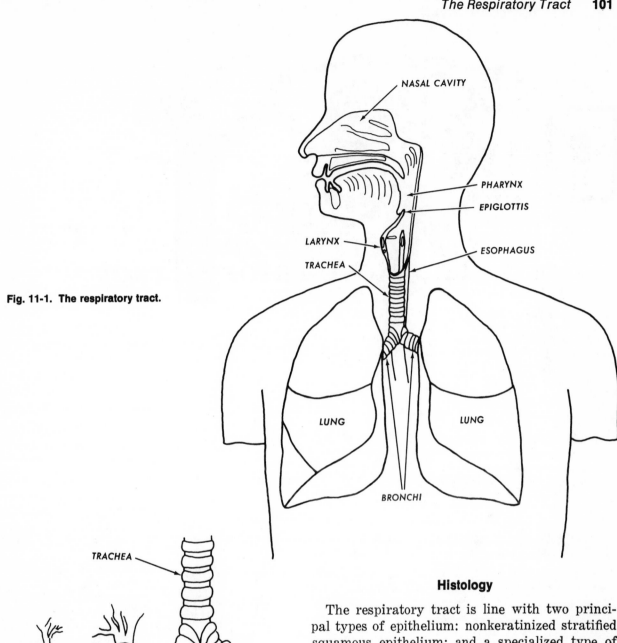

Fig. 11-1. The respiratory tract.

Fig. 11-2. Different branches of the airway.

Histology

The respiratory tract is line with two principal types of epithelium: nonkeratinized stratified squamous epithelium; and a specialized type of epithelium, known as respiratory epithelium. The squamous stratified epithelium is found primarily in the nasal cavity, mouth, and parts of the pharynx and larynx. The rest of the tract is lined with the respiratory epithelium.

Fig. 11-4 shows the structure of the respiratory epithelium. Note the predominance of ciliated cells. These cilia beat rhythmically toward the throat and mouth, and serve to remove foreign matter from the tract. The goblet cells, as shown in the figure, produce the substance known as mucus. Closer to the basement membrane are cells known as basal cells which play a role in the regeneration of the epithelium by replacing damaged or dead ciliated or goblet cells. Note that the

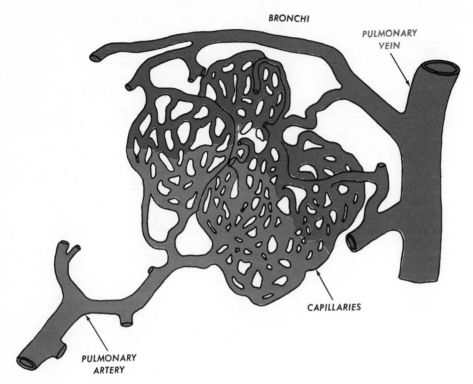

BRONCHI

PULMONARY VEIN

Fig. 11-3. Pulmonary circulation.

CAPILLARIES

PULMONARY ARTERY

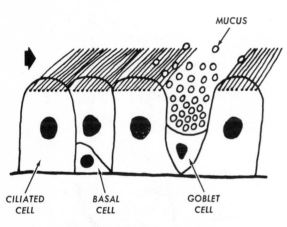

MUCUS

CILIATED CELL **BASAL CELL** **GOBLET CELL**

Fig. 11-4. Respiratory epithelium.

different heights of the various nuclei give the characteristic pseudo-stratified appearance.

Structure of Respiratory Tract Segments

Fig. 11-5 depicts a cross sectional view of the trachea. It consists of about twenty horseshoe-shaped rings of cartilage, joined at their openings by the tracheal muscle. The trachea is lined with respiratory epithelium, separated from the lamina propria by a basement membrane.

The section of the bronchus found outside of the lung has a structure similar to the trachea but of a smaller diameter. Inside the lung, the bronchus

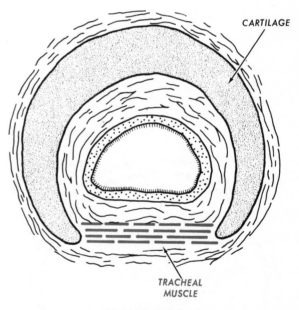

CARTILAGE

TRACHEAL MUSCLE

Fig. 11-5. Trachea.

appears more like that shown in Fig. 11-6. The cartilage rings have been replaced by plates of cartilage connected by smooth muscle. The mucous membrane too is made up of respiratory epithelium. As the bronchus divides and subdivides into increasingly smaller tubes, it contains less and less cartilage. The smaller subdivisions of the bronchus are known as bronchioles.

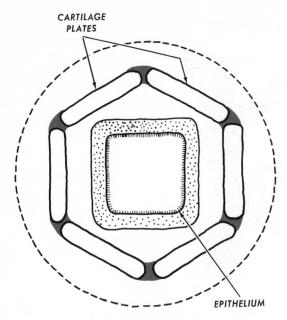

Fig. 11-6. Bronchus.

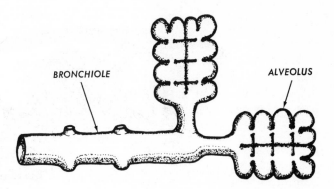

Fig. 11-8. Bronchiole and alveoli.

Although bronchioles contain no cartilage, they have more connective and elastic tissue, as shown in Fig. 11-7. As the bronchioles divide into smaller branches, they lose their goblet cells. This may be because mucus from these cells could clog the tiny passages. As the bronchioles get smaller and smaller, they also lose their ciliated cells, and are lined with cuboid or low columnar epithelium. Bronchioles eventually lead to alveolar ducts which lead to clusters of alveoli (Fig. 11-8).

The alveoli are grouped so tightly together that they share walls. These alveolar pockets are lined with squamous epithelium and are coated with a liquid layer called the surfactant, which reduces surface tension so that the lungs can expand easily. Fig. 11-9 shows an alveolar wall. Note the squamous epithelium made up of pulmonary surface epithelial cells, and the surfactant. The blood

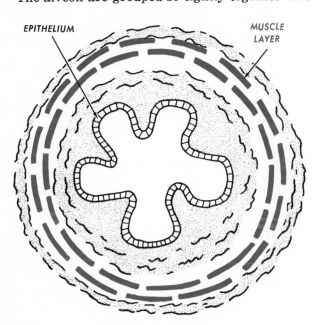

Fig. 11-7. Bronchiole.

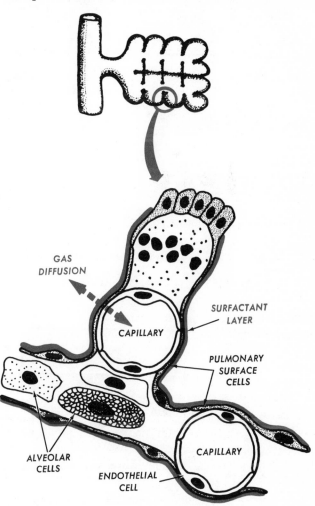

Fig. 11-9. Alveolus.

vessels are so thin at this point that they consist of a single layer of endothelium. Gas diffusing across the air-blood barrier must pass across the surfactant, pulmonary surface cell, basement membranes (fused together) and endothelial cell. Other cells found in the alveolar wall are known as alveolar cells, which are capable both of ame-boid movement and phagocytosis (the ability of a cell to engulf a particle of another cell). These cells can pick up lipids from the blood or particles of foreign matter. These cells then pass into the alveolar cavity and are pushed along the respiratory tract and into the throat where they can be found in the sputum.

Review Questions

1. Label the following parts in the illustration below: Trachea, bronchus, alveoli, pulmonary artery, capillaries, vein.

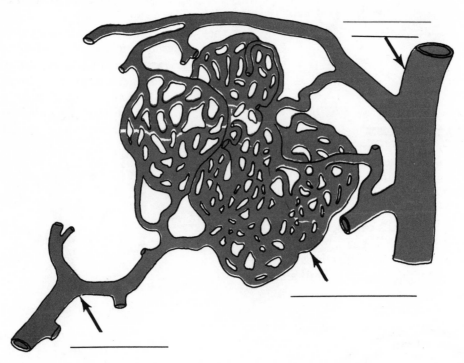

2. Label the following parts in the illustration below: Ciliated cell, basal cell, goblet cell, basement membrane.

3. Match the following labels with the illustrations below:
   a) alveolus
   b) bronchus
   c) trachea
   d) bronchiole

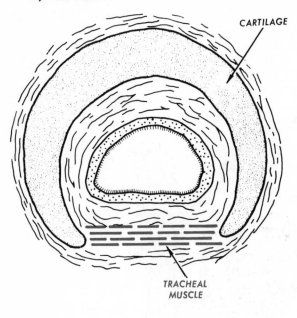

CARTILAGE

TRACHEAL
MUSCLE

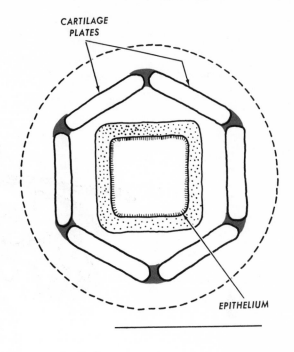

CARTILAGE
PLATES

EPITHELIUM

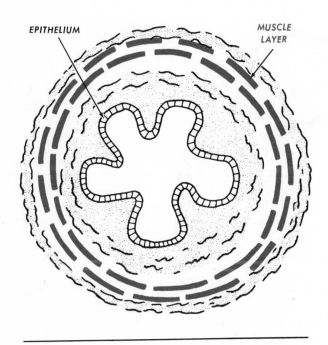

EPITHELIUM

MUSCLE
LAYER

4. Label the following parts in the illustration below: Surfactant layer, endothelial cells, pulmonary surface cells, capillaries, alveolar cells

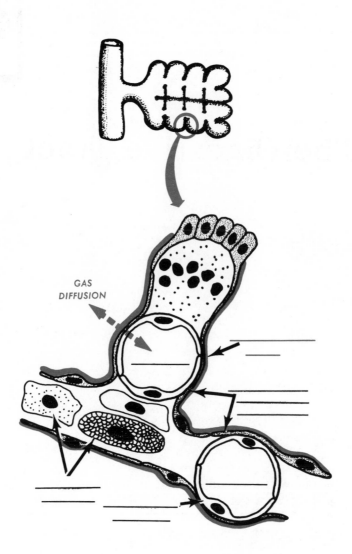

chapter 12

The Digestive Tract

One of the organ systems susceptible to cancer is the digestive tract. The parts of the tract which are of interest to the cytologist are the mouth, esophagus, stomach, duodenum (which receives secretions from the pancreas and liver) and the colon. Other regions are also susceptible to disease, but are not as accessible to cytologic examination.

As mentioned in the section dealing with the respiratory system, the mouth is lined with a non-keratinized stratified squamous epithelium, while the pharynx is lined in some areas with this type of epithelium and in other areas with respiratory epithelium.

Fig. 12-1 is a schematic representation of the gastrointestinal tract. The Latin word for stomach is *gaster*, so the adjective gastric refers to this organ. Food passes from the mouth, through the pharynx, and is propelled down the esophagus into the stomach. Here, it is mixed with gastric juices and is passed into the intestinal tract, where it is digested and the nutrients are absorbed. At the end of the intestinal tract, the food is expelled from the body through the anus.

Structure

Fig. 12-2 shows the basic structure of the entire digestive tract, from the esophagus to the anus. Note that it is made up of four main layers: the mucosa, or lining; the submucosa; the muscle layer; and the outer covering, or serosa. Each segment of the tract has its own characteristic variations of this basic design. For the sake of

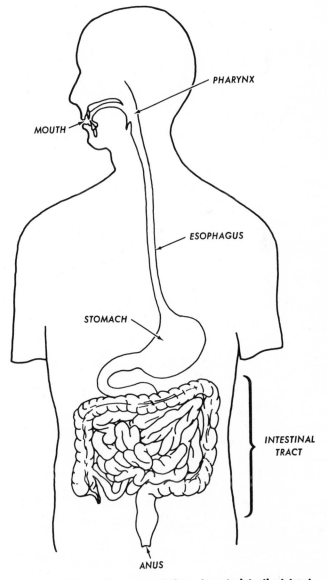

Fig. 12-1. Schematic representation of gastrointestinal tract.

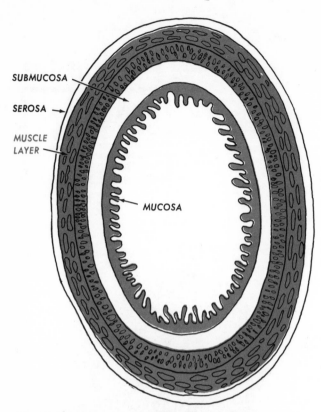

SUBMUCOSA

SEROSA

MUSCLE
LAYER

MUCOSA

Fig. 12-2. General structure of the entire digestive tract.

The Esophagus

The esophagus is a muscular tube which extends from the pharynx to the stomach. It passes near numerous organs, such as the trachea, and is susceptible to cancer invasion from neighboring organs. It also can develop its own tumors which may spread to other organs. The importance of early cancer detection is obvious.

The action referred to as swallowing is due to waves of contraction which force the ball of food into the stomach, even when the subject is placed in an upside-down position. Although the ball of food is already moistened with saliva from the mouth, the esophagus has glands which aid in this lubrication process.

The mucous membrane of the esophagus is lined with a tough, non-keratinized stratified squamous epithelium which is thick enough to resist abrasion from rough foods. Fig. 12-4 shows a cross section of this epithelium, with the characteristic fingers of lamina propria protruding into the epithelial layer. It is not very different from the epithelium which lines other organs, such as the mouth or the vagina.

this course, we are mainly interested in the mucosa. It consists of a surface epithelium, a lamina propria, and a muscularis mucosa, which is a less developed muscle layer found in the mucosa itself (Fig. 12-3). We now will discuss the digestive tract, section by section.

The Stomach

The stomach is a fibromuscular bag which stores food and mixes it with gastric juice. Fig. 12-5 shows the two zones distinguished by the histologist, the fundus of the stomach and the pylorus of the stomach. The epithelium of the

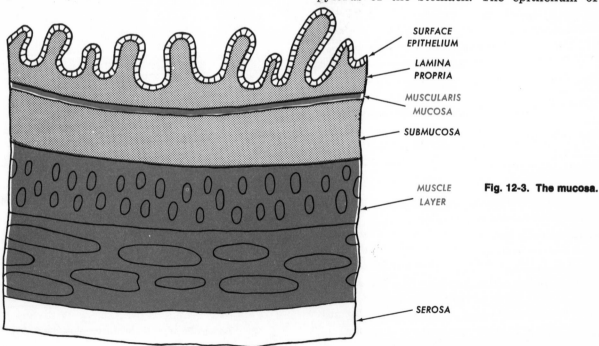

SURFACE
EPITHELIUM

LAMINA
PROPRIA

MUSCULARIS
MUCOSA

SUBMUCOSA

MUSCLE
LAYER

Fig. 12-3. The mucosa.

SEROSA

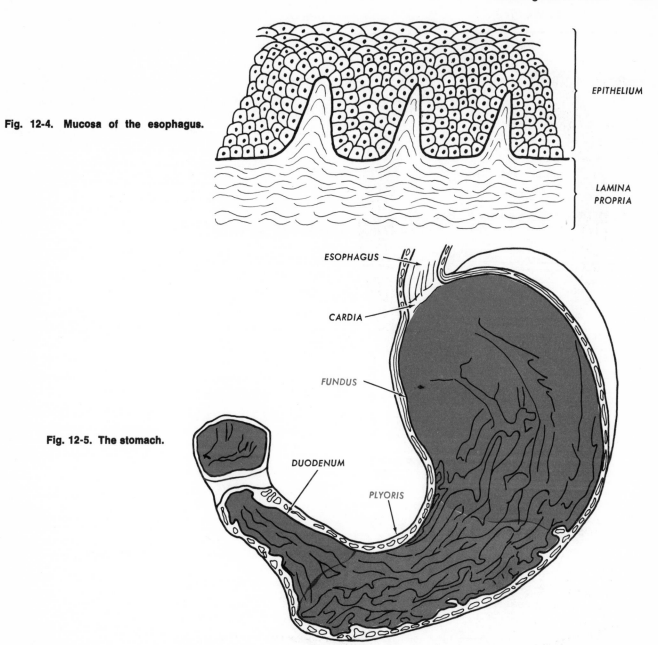

Fig. 12-4. Mucosa of the esophagus.

EPITHELIUM

LAMINA PROPRIA

ESOPHAGUS

CARDIA

FUNDUS

Fig. 12-5. The stomach.

DUODENUM

PLYORIS

esophagus changes suddenly at the cardia of the stomach, where it assumes the appearance characteristic of the fundus. The epithelium changes its appearance again in the pyloric region. In the stomach mucosa, the epithelium consists of two zones, the gastric pits, and the glands, which vary between the fundic and pyloric regions.

Fig. 12-6 shows the abrupt change in the epithelium of the esophagus and the fundus. Note that the epithelium of the fundus forms distinct pits in the surface which lead down to the gastric glands. These crypts are wide and shallow, as compared to those of the pyloris. The glands are straight tubes which are made up of three differ-

ent cell types (Fig. 12-7). The surface epithelium consists mainly of columnar cells with clear regions which contain mucus. This mucus is secreted as protection against the powerful stomach acids. In the glands, the predominant cell type is known as the chief cell, which secretes digestive enzymes such as pepsin. The neck region of the gland is lined with so-called mucus neck cells which, like the surface cells, secrete mucus. The bulging parietal cells produce hydrochloric acid. It is worth mentioning in addition that in the cardiac region of the fundus of the stomach, still another type of gland, known as the cardiac gland, also secretes mucus.

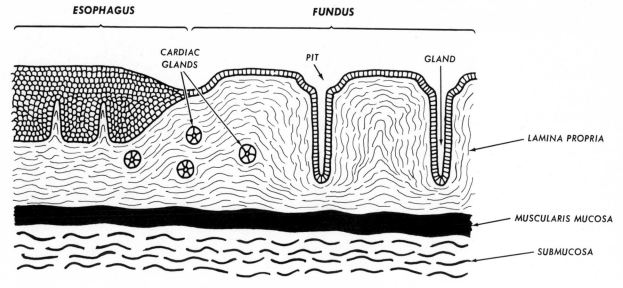

Fig. 12-6. Function of esophagus and fundus.

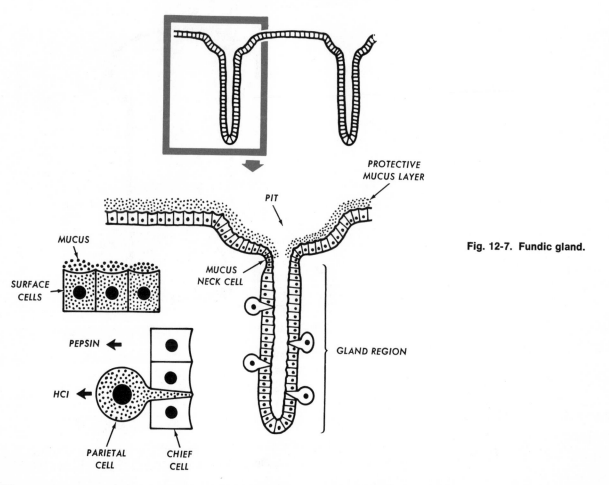

Fig. 12-7. Fundic gland.

Fig. 12-8 compares the appearance of the fundic mucosa to that of the pylorus. Note that there are major differences in the pits: in the pylorus they are narrow and deep, while in the fundus they are wide and shallow. The glands, too, are different. Instead of the straight fundic glands, we see the coiled pyloric glands which appear in cross section in the diagram. The pits are lined the same way as those of the fundus, but the glands are made up of only one cell type.

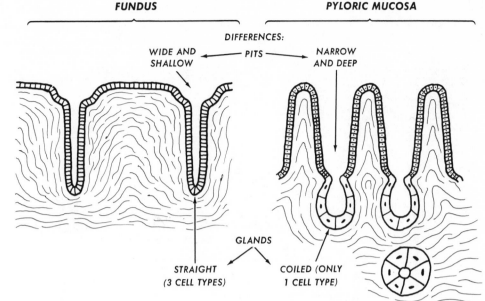

Fig. 12-8. Function of fundus and pylorus.

The Small Intestine

The small intestine actually is about 10 to 20 feet long, and has several functions. It serves to transport food from the stomach to the large intestine, food which is digested along the way by digestive enzymes. The small intestine also absorbs the nutrients produced by this digestion, and secretes certain digestive hormones into the blood stream to control this process. Since absorption is best accomplished by a greater surface area, the small intestine is organized in a special way. The entire mucosa is arranged in spiral folds along the interior of the tube (Fig. 12-9). Each fold is lined with villi, or fingers, which project into the lumen (Fig. 12-10). The villus itself, as shown in Fig. 12-11, is made up of cells which also have microvilli, again increasing the surface area. The small intestine has a total surface area of 20 to 40 square meters (65 to 95 sq. ft.).

Just as the stomach mucosa is divided into pits and glands, so the lining of the small intestine is organized into villi and crypts of Lieberkuhn (Fig. 12-12). Beneath the mucosa are found the glands of Brünner (Fig. 12-13).

The Duodenum

The duodenum is a C-shaped section of the small intestine which lies between the stomach and the rest of the small intestine. The second section of the duodenum, as shown in Fig. 12-14, receives secretions from the liver and pancreas through ducts. The pancreas, aside from its endo-

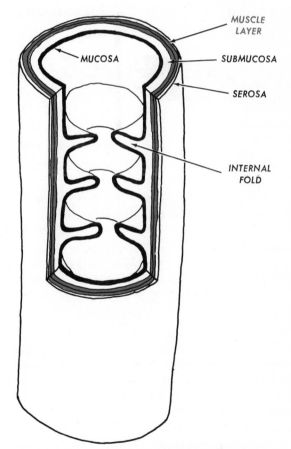

Fig. 12-9. The internal folds of the small intestine.

crine function of blood sugar control, secretes a large quantity of digestive enzymes, while the liver secretes bile. These ducts leading from the liver and pancreas are lined with a single layer of cylindrical epithelium.

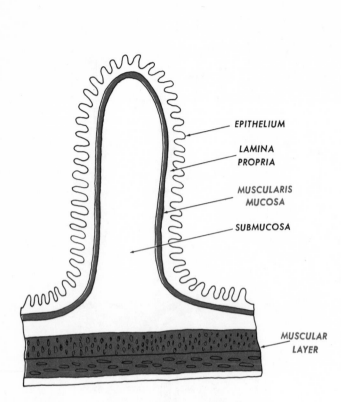

Fig. 12-10. The intestinal villi.

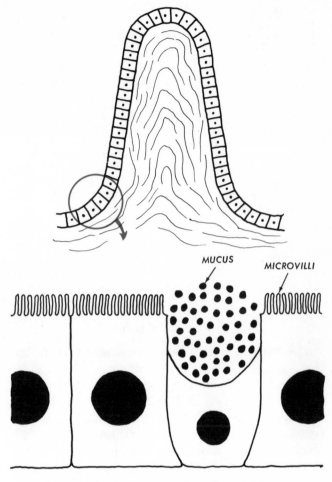

Fig. 12-11. Intestinal epithelium.

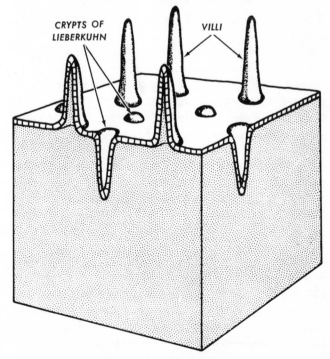

Fig. 12-12. Intestinal villi and crypts.

The epithelium of the duodenum consists of three cell types (Fig. 12-15). The most numerous cell is the columnar cell with striated border. This striated border consists of microvilli, very well suited to absorption. The goblet cells are less numerous. They contain fine grains of mucus. The third cell type, the argentaffin cell, is named for its affinity for salts of silver (hence the word "argent").

Fig. 12-16 shows a magnified view of the glands or crypts of Lieberkuhn. There are four cell types present: the columnar cell with striated border, goblet cell, argentaffin cell, and the cells of Paneth. Although no one knows their precise function, the cells of Paneth are believed to be enzyme secreting cells.

The cells of the small intestine and the stomach also secrete a number of hormones which regulate the digestive process. The cells lining the stomach and the intestine are exposed to powerful enzymes and acid and must be replaced at a rapid rate. It is necessary for the lining to be completely replaced every few days.

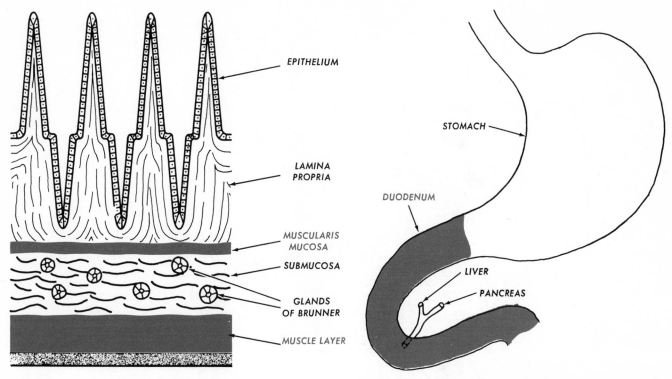

Fig. 12-13. Duodenal mucosa.

Fig. 12-14. The duodenum.

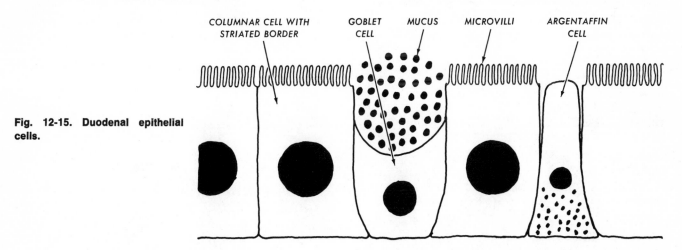

Fig. 12-15. Duodenal epithelial cells.

The Large Intestine

The large intestine, or colon, is about five feet long and leads to the anus. Its primary function is the absorption of water and electrolytes, although it also secretes mucus to lubricate the passage of the feces which are being formed. The large intestine has a population of bacteria, known as the intestinal flora, which produce vitamins.

The mucosa of the colon is smooth. There are no villi or folds, only crypts. Fig. 12-17 shows a cross-section of the mucosa. The surface is lined with two cell types; the columnar cells with a striated border, and the goblet cells. The crypts, like those of the small intestine, are made up of columnar cells and goblet cells, but no Paneth cells.

Finally, the mucosa of the colon meets with the anal mucosa. The epithelium changes from a columnar to a non-keratinized stratified squamons epithelium and then to a keratinized stratified squamous epithelium which is the skin of the body.

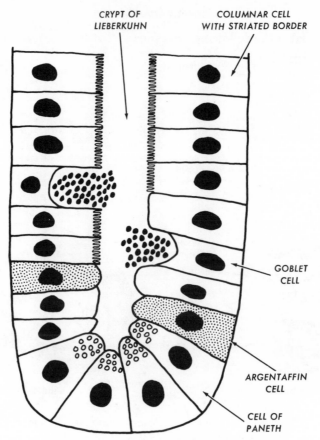

CRYPT OF
LIEBERKUHN

COLUMNAR CELL
WITH STRIATED BORDER

GOBLET
CELL

ARGENTAFFIN
CELL

CELL OF
PANETH

Fig. 12-16. Detail of crypt of Lieberkuhn.

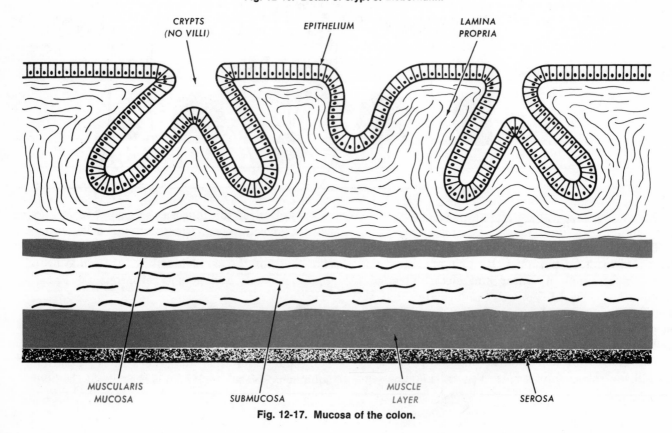

CRYPTS
(NO VILLI)

EPITHELIUM

LAMINA
PROPRIA

MUSCULARIS
MUCOSA

SUBMUCOSA

MUSCLE
LAYER

SEROSA

Fig. 12-17. Mucosa of the colon.

Review Questions

1. In the illustration below, label the following: Surface epithelium, lamina propria, muscularis mucosa, submucosa, muscle layer, serosa

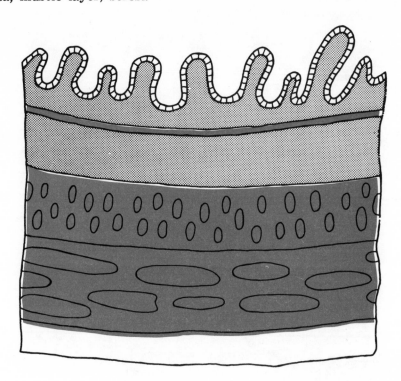

2. In the illustration below, label the following: Esophagus, cardia, fundus, pylorus, duodenum

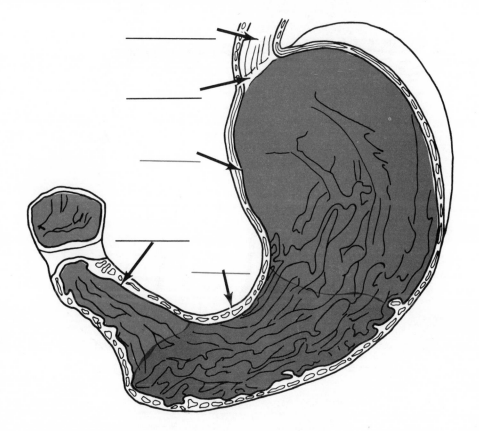

3. Match the following labels to the illustrations shown below.
    a) pylorus
    b) esophagus
    c) duodenum
    d) colon
    e) fundus

(See additional figures on Page 117.)

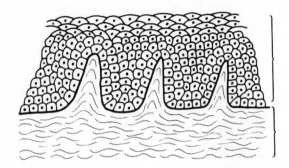

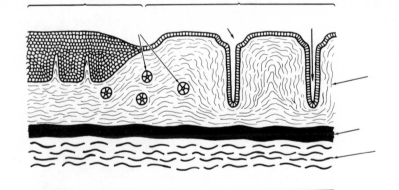

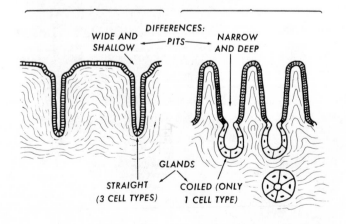

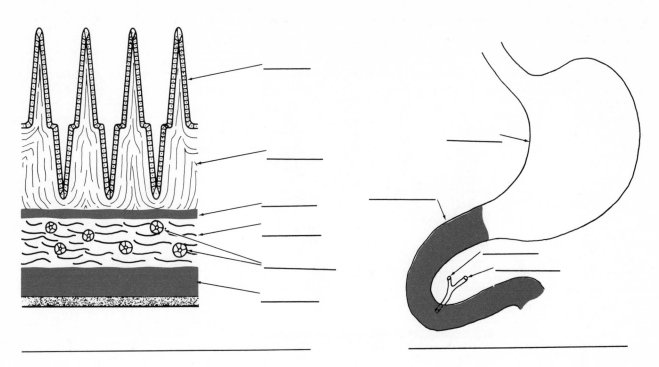

4. Label the following parts of fundic gland in the illustration below: Pit, mucus layer, mucus neck cells, parietal cells, chief cells

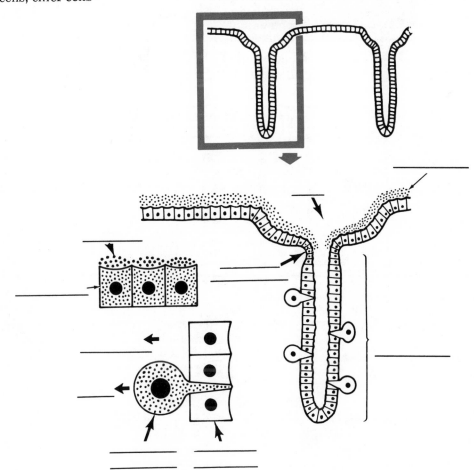

5. Name substances secreted by:

           chief cell _____

           parietal cell _____

6. In the illustration below, label the following: Epithelium, lamina propria, muscularis mucosa, submucosa, glands of Brünner, muscle layer

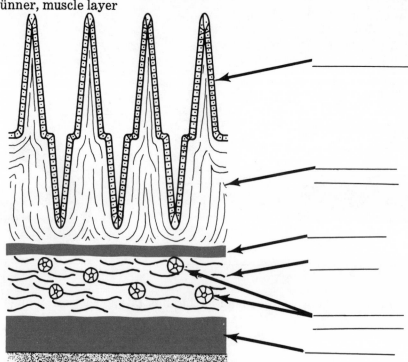

7. In the illustration below, label the following parts of the crypt of Lieberkuhn: Columnar cell with striated border, goblet cell, argentaffin cell, cell of Paneth

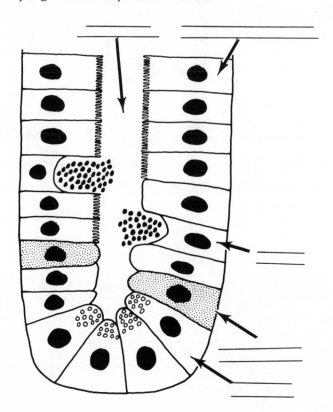

8. In the illustration below, label the following: Crypts, epithelium, lamina propria, muscularis mucosa, submucosa, muscle layer, serosa

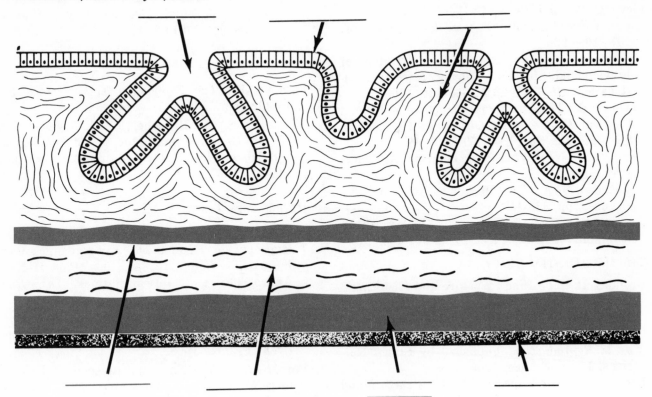

Recommended Reading

GENERAL REFERENCES

Brachet, J., and Mirsky, A. E. (Eds.): *The Cell.* New York: Academic Press, 1959-1961.

Dippell, R. V.: Ultrastructure of Cells in Relation to Function, in *This Is Life.* Johnson, W. H., and Steere, W. C. (Eds.). New York: Holt, Rinehart and Winston, Inc. 1969.

Giese, Arthur Charles: *Cell Physiology.* (2nd ed.) Philadelphia: W. B. Saunders Co., 1963.

Neutra, M., and Leblond, C. P.: The Golgi Apparatus. *Scientific American,* **220,** 1969.

Swanson, C. P.: *The Cell,* (3rd ed.) Englewood Cliffs, N.J.: Prentice-Hall Inc., 1969.

GENETICS

Hershowitz, I. H.: *Basic Principles of Molecular Genetics.* Boston: Little, Brown & Co., 1967.

Kornberg, A.: Biosynthesis of DNA. *Scientific American,* **219**:64, 1968.

METABOLISM

Lehninger, A. L.: *Biochemistry: The Molecular Basis of Cell Structure and Function.* New York: W. A. Benjamin, Inc., 1970.

UROGENITAL

Brachet, J., and Mirsky, A. E. (Eds.): *The Cell.* New York: Academic Press, 1959-1961.

Leeson, C. R., and Leeson, T. S.: *Histology.* Philadelphia: W. B. Saunders Co., 1966.

Myers, C. E., Bulger, R. E., Tisher, C. C., and Trump, B. F.: Human Renal Ultrastructure, *Laboratory Investigation,* **15**: 1921-1950, 1966.

Pitts, R. F.: *Physiology of the Kidney and Body Fluids.* Chicago: Year Book Medical Publishers, 1963.

Ultrastructure of the Kidney. New York: Academic Press, 1967.

FEMALE GENITAL TRACT

Barth, R.: Endocrinology of the Human Menstrual Cycle. Opinions and Hypothesis. *Vitamins & Hormones,* **25**:123-135, 1967.

Chiazze, L., Jr., Bayer, F. T., Macisco, J. J., Jr., Parker, M. P., and Dufby, B. J.: The Length and Variability of the Human Menstrual Cycle, *J.A.M.A.,* **203:6,** 377-380, 1968.

Lang, W. R., and Sponte, G. E.: Epithelial Regeneration of the Human Uterine Cervix. *Am. J. Ob-Gyn.,* **91:5,** 657-664, 1965.

Lloyd, C. W.: *Textbook of Endocrinology*. The Ovaries. Philadelphia: W. B. Saunders & Co., 1968, pp. 449-536.

Lloyd, C. W., and Weizz, J.: Some Aspects of Reproductive Physiology. *Ann. Rev. Phys.* **28:** 267-310, 1966.

McLennon, C. E., and Rydell, A. M.: Extent of Endometrial Shedding During Normal Menstruation. *Ob-Gyn.*, 26:5, 605-621, 1965.

Obeblad, E.: The Functional Structure of Human Cervical Mucus, *Acta, Obstetrica Gynecologia Scandinavica*, Suppl. #1, 59-79, 1968.

The Ovary. New York and London: Academic Press, 1962.

—Asdell, S. A.: The Mechanism of Ovulation, Vol. 2, pp. 435-449.

—Jones, I. C., and Ball, J. N.: Ovarian Pituitary Relationships, Vol. 1, pp. 361-434.

THE MALE GENITAL TRACT

Brandes, D.: The Fine Structure and Histochemistry of Prostatic Glands in Relation to Sex Hormones. *Int. Rev. Cytol.*, **20:**206-276, 1966.

Girod, C., and Czyba, J. C.: *Cours sur la Biologie de la Reproduction*. Published by Simep Editions, Lyon, France, 1969.

Lacy, D.: Certain Aspects of Testis Structure and Function, *Brit. Med. Bull.*, **18:**205, 1962.

Nilsson, S.: The Human Seminal Vesicle. A Morphagenetic and Gross Anatomic Study With Special Regard to Changes Due to Age and to Prostatic Adenoma. *Acta Chirurgica Scandinavia*, (Suppl. 296) 96 pp, 1962.

THE DIGESTIVE TRACT

Kramer, P.: The Esophagus. *Gastroenterology*, **49:**439-463, 1965.

Melnyk, C. S., and Braucher, R. E.: Regeneration of the Human Colon Mucosa. Morphological and Histological Study. *Gastroenterology*, **52:** 985-997, 1967.

RESPIRATORY TRACT

Bertalanfly, F. D.: Respiratory Tissue: Structure, Hypsophysio Histophysio, Cytodynamics—Part I. *Int. Rev. Cytol.*, **16:**234, 1964. Part II—New Approaches and Interpretations. *Int. Rev. Cytol.*, **17:**214, 1964.

Czyba, J. C., and Girod, J. C., *Cours D'Histologie et Embryologie*, Published by Simep Editions, Lyon, France, 1968.

Low, F. N.: The Pulmonary Alveolae of Laboratory Animals and Man. *Anat. Rec.* **117:**241, 1953.

Index

Cell: biochemistry, 23
  energy, 23
  enzymes, 23

Cell: genetics, 26
  DNA, 26
  genetic code (RNA), 28
  nucleic acids, 26

Cell: physiology, 18
  membrane transport, 18
    active, 19
    passive, 19
  vacuole transport, 20
    exocytosis, 20
    ingestion of liquid materials, 20
    phagocytosis, 20
    pinocytosis, 20

Cell: reproduction and tissue formation, 32
  centrioles, 33
  chromosome number, 32
  cycle, 35
  epithelium, 36
  mitosis, 33
  organization into tissues, 35
  sex chromatin, 35

Cell: structure, 9
  cytoplasm, 10
  cytoplasmic inclusions, 14

endoplasmic reticulum, 10
Golgi apparatus, 14
lysosomes, 12
membrane, 9
mitochondria, 11
nucleus, 15
ribosomes, 12
Digestive tract, 107
  duodenum, 111
  esophagus, 105
  large intestine, 113
  small intestine, 111
  stomach, 108
  structure, 107
Female genital tract: fallopian tubes, and uterus, 59
  during endocrine cycle, 60
  menstruation, 63
  myometrium, 61
  uterine contractions, 63
  uterus, physiology of, 62
Female genital tract: ovaries, 40
  drug controlled ovulation, 45
  endocrine control of ovulation, 45
  estrogens, 45
  follicular development, 42
  menarche, 49
  menopause, 50
  ovary, 40

progesterone, 45
Female genital tract: vagina, 71
  endocervical canal, 80
  lactic acid and vaginal pH, 72
  section of vaginal epithelium, 72
  vagina, 71
    epithelium cycle, 75
    flora, 72
    mucosa, 72
    sex activity, 72
    smear, 75, 79
Male genital tract, 91
  ejaculation, 96
  ejaculatory duct, 92
  epididymis, 91
  prostate gland, 96
  semen, 91
  testicles, 91
  urethra, 93
  vas deferens (ductus deferens), 92
Respiratory tract, 100
  histology, 101
  structure, 100, 102
Urinary tract, 83
  bladder, 87
  excretory passages, 85
  kidneys, 83
  renal pelvis and ureters, 86
  urethra, 87